Manuals in Biology

Zoonoses of Primates

Manuals in Biology

Editors: Richard Carrington and

Dr L. Harrison Matthews, F.R.S.

Zoonoses of Primates

The Epidemiology and Ecology of Simian Diseases in Relation to Man

Richard Fiennes

M.A. (Cantab) Nat.Sci., M.R.C.V.S. (Edin.)

* * *

Cornell University Press

ITHACA, NEW YORK

CORNELL UNIVERSITY PRESS
1967

Library of Congress Catalog Card Number 67–20652

PRINTED IN GREAT BRITAIN

Contents

List of Tables

Preface

ZOONOSES OF PRIMATES is the first volume of *Manuals in Biology*, a new series of monographs by world authorities intended for research workers and students in the sciences. The Manuals will be strictly professional and, although they will make no attempt at general appeal, they will interest those working in disciplines allied to their special subjects. The volumes will vary in length and amount of illustration according to the needs of the subject and the author's treatment of it.

Zoonoses are the diseases of animals transmissible to man, in whom they often produce symptoms much more serious than they do in their non-human hosts. The latter are carriers that may have achieved an equilibrium with their parasites so that no adverse symptoms are seen, but when transmission to man occurs the resulting disease may produce grave consequences. Among non-human primates the simian carriers of yellow fever and B-virus are well-known reservoirs of such diseases.

Apart from the risk of human infection from wild populations of primates, the very large numbers of monkeys and apes now used for scientific and medical research, and for the production of prophylactic and other sera, are a potential source of zoonoses because few of the animals are bred in captivity and the majority are captured in the wild. During the last decade several tragic occurrences have emphasised the importance of research on the subject in order to safeguard those working with primates.

The literature of the zoonoses of the primates is widely scattered in a host of journals; the author of this volume has been at much pains to gather all of importance that is known about them. This book therefore gives a firm foundation on which to build further research.

L. HARRISON MATTHEWS
RICHARD CARRINGTON

Foreword

The Concept and Scope of the Study of Zoonoses Amongst Primates

Compilation of *The Zoonoses of Primates*, when requested by the Food and Agriculture Organisation, appeared at first sight to be a simple matter of recording and discussing those diseases which can be transmitted from the pro-simian and simian members of the primate group to that aberrant species *Homo sapiens*. The term 'Zoonoses' used within the one class of primates appeared confusing, but on reflection was justifiable in the sense defined above. The task, however, proved to be beset with many unforeseen complexities, which involved the author in extensive researches into literature, much of which was not easily traced or obtainable. To produce a catalogue of those diseases which are or can be shared by monkeys, apes and man is admittedly not a difficult task. A guide to these has been published by Appleby *et al.* (1963); and much of the information can be acquired from such works as Ruch (1959) *Diseases of Laboratory Primates*, and from the New York Academy of Sciences (1969), *Care and Maintenance of the Research Monkey*. These sources would be adequate for the assembly of a laboratory handbook, which could be useful in warning those who handle primates of the dangers to human health.

Nevertheless, the dangers of acquiring primate diseases by direct contact, though of great consequence to the small section of the community which may handle these animals, is of far lesser importance than the epidemiological threat which primates pose for human communities living in the same neighbourhood. The consequences of disease transmission from one to the other, as witnessed by yellow fever epidemics and many other instances below, are so important that any comprehensive work on primate zoonoses must for the greater part be occupied with them. To understand these dangers it is necessary to study both the phylogenetic and the ecological relationships between man and other members of the primate group. Since the ecological relationships are indirect, a study of the bionomics of arthropods and other vectors by which disease could be transmitted is clearly needed. An investigation must be made of the disease cycles involving simian and human primates with other susceptible animals, including tree and air-borne species, such as birds and bats, and ground-living species, which might come in contact with man. The information for such a comprehensive study is fragmentary; however, an attempt is made here to present what is known, and to indicate the great importance to human health that should be attached to the acquisition of further knowledge on the subject.

The term 'zoonosis' (*ΖΩΟΝ*=living thing; *ΝΟΣΟΣ*, or *ΝΟΣΗΜΑ*=disease, plague) has been largely misunderstood and misinterpreted by British authors, who have introduced a number of unjustified and mistaken terms to expand the meaning to cover various different disease relationships between animals and man. Both Hoare (1962) and Nelson (1960) attribute the introduction of the term to Virchow, the father of modern pathology. Neither author, however, gave reference to Virchow's substantive paper, and it appeared advisable to discover the original meaning of zoonosis. Accordingly, an extensive investigation into the origins of this term was made on behalf of the author by Miss E. Von Bernuth, with surprising results.

The term zoonosis appears to have originated in Germany and to have been in common use by the middle of the nineteenth century, and probably had been used for many generations prior to that. The meaning is quite simply that of the Greek words, namely a disease of animals—as opposed to disease of man. It was used in this sense, *and in no other*, by Virchow, who thus had no hand in the introduction of a new term with a specialised meaning. It appears to have been used by German physicians in describing certain diseases in patients who came in contact with animals and appeared to acquire diseases from them. Thus, one can understand how a doctor, seeing perhaps a butcher with a 'malignant pustule' would say to him:

> *Das ist keine Krankheit der Menschen, das ist eine Krankheit der Tiere, eine Zoonosis.*

Thus, in the course of time, the term came to have a secondary meaning implying a human disease which was acquired from an animal source, and which was not normally transmitted from man to man.

In this sense it is defined by Mayne (1860) in his *Expository Lexicon of the Terms Ancient and Modern in Medical and General Science*, which gives:

> Zoonŏsema zoon = an animal, nosēma = a disease) zool. term by C. H. Schultze for diseases (n., pl.) in herds of cattle. French analogy *zöonosème*, m. m. Germ. syn. *die Krankheit mit tierischem Heerde.*

Miss Von Bernuth points out that Mayne is guilty of a mistranslation, since the German word *Heerde* does not mean herd, but hearth or origin, thus the meaning which Mayne should have given would be 'diseases of animal origin'. Mayne appears to have been somewhat uncertain about the uses and origin of this term, since it was omitted from his *Medical Vocabulary* (1862), and in the revised edition (1875).

Probstmayer (1863) in his *Dictionary of Veterinary Medicine* gives the meaning of zoonoses as:

> Zoonoses are firstly original animal diseases; secondly diseases of man which can be transmitted to him by means of a contagium from animals.

Thus, in the 1860s the term had reached the medical and veterinary dictionaries and was recognised to carry the double meaning, either a disease of animals, or an animal disease which could be transmitted to human beings. In this sense it appears today in Henderson (1963) *Dictionary of Scientific Terms*, which defines it as follows:

> Disease of animals; animal disease transmitted to man.

This dictionary also refers to the term 'zoosis', meaning any disease produced by animals. Dorland (1949) *Medical Dictionary* defines the term as a disease of animals that may secondarily be transmitted to man.

The Joint WHO/FAO Expert Committee on Zoonoses (1959) defined them as:

> Those diseases and infections naturally transmitted between vertebrate animals and man.

This definition is unexceptionable, both etymologically and for convenience in practical usage. It would be permissible, additionally, to accept 'anthroponoses' as diseases primarily of man, which could be transmitted to other animal species. Such a term would be occasionally useful, and the meaning is clear. Many other terms should be discarded as etymologically incorrect and of doubtful value. For instance Nelson introduces the term 'anthropozoonoses', meaning 'affections of man naturally acquired from other vertebrates where the maintenance host is animal and man is an incidental host ...'. In the inverse sense he introduces 'zoo-anthroponoses'. He is, however, disregarding the fact that the terms are inverted as defined by the original meaning of zoonosis, an animal disease; which must make his anthropozoonosis mean a man-disease, transmissible to animals – the exact opposite of his own definition.

The best review of the etymology of zoonoses is given by Koegel (1951) who correctly defines the term and stresses its relationship with the terms 'nosology' or 'nosography', which were used in this country in the past, as alternatives for pathology, and are still so used in a number of Continental countries including the USSR.

Further discussion of the meaning of the term 'zoonosis' is unnecessary, except to reject all other terms except 'anthroponoses'. These two terms, used in a broad sense, cover all required meanings and much of the additional terminology which has been introduced is absurd. For instance, on the basis of etymology, a 'platyhelminthozoonosis' would mean a disease of Platyhelminths, transmissible to man, but *not* an animal disease caused by Platyhelminths from which man could also suffer! The present work indicates those conditions of bacterial, viral, helminthic and protozoal origin which can be shared by man with the pro-simian and simian members of the primates; a group to which man also belongs. In this sense the distinction between zoonoses and anthroponoses becomes somewhat academic, although it may sometimes be useful. The familial relationships between man and other primates, however, are of first class importance and by themselves make a fascinating study. Some parasites may infect both groups, causing little harm to either. Other parasites may infect one group but not the other, and yet others may be natural parasites in the one group and have grossly enhanced virulence for the other; in this way being responsible for the deadly dangers of infections which occasionally arise from handling monkeys. An example of such is the B-virus. Thus there are special problems involved in the study of primate zoonoses, which cannot be adequately undertaken without a review of the familial and ecological relationships of the two groups as discussed in the next two chapters.

It is not only in relation to the term zoonoses that etymological difficulties arise. Many authors, including Ruch, have discussed the problems of nomenclature which arise in the differentiation of man from other primate groups. Some authors speak of 'subhuman primates'; others of 'non-human primates'. All such terms are objectionable from one point of view or another, if only because they are clumsy. Innes (1965) has suggested the use of 'human primates' and 'simian primates'; in brief 'humans' and 'simians'. Provided one defines the word simian to cover also pro-simian where meant and intended, this suggestion offers the best solution to the problem, and this terminology is adopted throughout this work. In general the word simian will cover both pro-simian and simian species; where intended to cover prosimian only the word pro-simian will be used. Some people will object to the use of the words 'human' and 'simian' as nouns; this objection is technically correct, but the word human as an abbreviation for human being has surely been accepted into the English language. The word simian, derived from *simia*, Latin:

ape, according to the *Concise Oxford Dictionary*, can be adjective or noun. The author begs leave to doubt the correctness of the dictionary, but proposes to use the word as a noun when convenient nevertheless.

Part I General

1 The Ecology of Primate Zoonoses

Most works on zoonoses simply describe the diseases suffered by certain animals or groups of animals, which can also infect human beings. For instance, certain diseases may be attributed to domestic animals, such as the dog, and these can under certain circumstances be dangerous to human children or adults. An example of a disease, which has recently caused considerable concern, is that of '*Larva migrans*', derived from a dog infection caused by the nematode worm *Toxocara canis* and transmitted under rather unusual and peculiar conditions of life history. There also exists in the literature the record of a case of a child dying from poliomyelitis contracted from a budgerigar. Such descriptions of zoonoses, though valuable, are only applicable to the simian world in cases where direct contact between man and simians occurs through the latter being brought into captivity, or, in comparatively rare cases as with the temple monkeys in India, where the two groups to some extent share a common habitat. This more direct view of primate zoonoses is of obvious importance to persons who handle captive simians and are at risk. It is, however, of far less social importance than that which arises as a result of the transfer of some disease of monkeys or apes from the closed arboreal habitat which most species inhabit to the massed human communities which our society creates. That such dangers exist is evident from the story of yellow fever, and further possible instances will be referred to later. What might happen if some virus of a virulence comparable with that of myxomatosis for rabbits should suddenly appear in packed human communities? It is a measure of the backwardness of virology at present that it might take some months before a vaccine could be produced in sufficient quantities to protect threatened communities; meanwhile a death rate of 90% or more could have occurred.

These dangers are probably far more real and serious than is generally supposed, yet we find that the science of ecology, as applied to medicine and to disease transmission, is one which is neglected in both human and veterinary medical courses. Most of what is known has been collected by ecologists; that is by professional zoologists who have specialised in this branch of their work. Useful studies have been made by such workers as Charles Elton, Paul Errington, Frank Fenner and others. Very few accounts of this kind of work are available in any literature which is commonly consulted by doctors or veterinary surgeons. An excellent introduction to the subject of disease cycles is contained in Harris (1962) *The Problems of Laboratory Animal Diseases*; this work contains contributions of high quality by Paul Errington, Bohumir Rosicky, Frank Fenner, E. Sacquet, Howard A. Schneider and others. Sir Macfarlane Burnet (1962) also has given a fascinating study of disease ecology in his *Natural History of Infectious Disease*. In the book *Man, Nature and Disease* (Fiennes, 1964) this author also has attempted to present the background of disease as a 'natural' phenomenon. Thus there exists some literature from which the theory of disease in nature can be studied. It is not possible here to discuss it in great detail, although certain remarks must be made as to the special circumstances affecting the transmission and maintenance of disease amongst natural populations of primates in their arboreal habitat, and of ways in which there could be transference to human settlements.

Man's own ecological links are sufficiently complex. During prehistoric time before the invention of agriculture, when he existed as a nomadic hunter and food-gatherer, his main relationship with other groups of animals was evidently that of predator with prey. This association still survives in his relationships with domesticated animals, and the larval forms of his tapeworms exist in ox and pig. These relationships resemble those between Carnivores, such as jackals, foxes and other Canidae, with Herbivores such as rodents and sheep which are the intermediate hosts of their tapeworms. However, since man became a civilised animal living in permanent settlements, the picture has become extremely complicated. This is due to the development of new relationships between him and field rodents, some of which live solely in and around human habitations, and some of which seasonally migrate from the fields to human settlements at times of food scarcity. By this means complex links can be created in which man becomes implicated in alien disease cycles, endemic in wild rodent populations. These may at times create epidemic disease in the human species, involving indeed other species of animals in addition. Human habitations, too, have become the resort of numbers of scavenging animals, particularly birds, such as sparrows, starlings, pigeons and seagulls, which may well introduce and spread all manner of diseases from psittacosis to intestinal infections associated with *Salmonellae*.

In general, animals which share a habitat, as in the human instances cited, become tolerant of each other and only transmit disease to each other at a dangerous level when their numbers become excessive. Dangers arise because parasites, and pathogens, are increased in numbers *pari passu* with the increase in host numbers, and so increase infection of the soil, vegetation, tree bark and so on. Secondly, there is danger as arthropod vectors become more numerous and more heavily infected; thirdly, the stamina of the host species is reduced, and parasites and pathogens tend to become more virulent; fourthly, the host species itself may migrate from its natural habitat and come in contact with other species, from which it is usually separated. Conditions are thereby created by which diseases may appear and may be passed from one host to another. Pastures may be contaminated with anthrax, black-quarter, tetanus and other soil-living organisms, capable of causing disease in many host species. Those diseases of fungal origin, which do not normally spread from animal to man but are derived from contaminated vegetable material, may become more widespread.

Diseases of bacterial origin, the causal organisms of which are normally quiescent in the natural host, become active and, as happens with tularaemia during the lemming migrations, come to infect abnormal hosts such as dogs and man. After a population crash, arthropod vectors such as mosquitoes, ticks and tsetse-flies, which are denied their natural hosts, may attack unnatural hosts and thereby infect them with diseases of viral, protozoal or rickettsial origin. At the same time the migrating surplus of the host population may have come into contact with related species, in which its natural pathogens cause severe and often fatal disease. This eventually may lead to widespread epidemics through the contact cycles of the secondary host. In this way the great plague epidemics of history probably originated; the natural hosts of the plague bacillus, gerbils and susliks, came in contact with black rats; the gerbil fleas infected the rats; the rats infested human habitations and the plague spread.

Factors which lead to the transmission of disease from one species to another, and which are associated primarily with the habitat, thus arise from two main sources: firstly, contamination of the habitat by reason of excessive numbers; secondly, unusual contact between species, which may be direct – or indirect through intermediate hosts.

The factors which determine host specificity in parasites and pathogens are not understood, although obviously the possibility that a species can transmit its parasites to man is an important factor in the zoonoses. Host and parasite become so adapted to each other by evolutionary means that mutual tolerance becomes increased, and in normal circumstances the host is not unduly harmed by its parasites. In this respect, also, the situation is entirely different from that of man and his domestic animals, because in civilised communities, instead of being tolerated, parasites are eliminated if possible. In the feral state in certain situations animals may harbour a tremendous host of normally harmless symbiotes, which in adverse circumstances may become pathogenic. These include intestinal pathogens (protozoal, bacterial or viral); blood parasites; many types of Platyhelminths: indeed so many potentially pathogenic organisms as to bewilder and confuse the medical or veterinary man who is not normally acquainted with such a situation. A typical example is shown by South American primates from the unhealthy Amazon Basin, which appear to be so adjusted to their natural habitat and its hazards that they can survive there, but which are difficult to keep alive when removed to captive conditions because they are then overwhelmed by their natural parasites.

Parasites, once adapted to specific hosts, tend to lose the power of infecting other hosts, but if they do so they often cause a violent reaction to their presence, resulting in severe disease, the end point of which is the death of the host, or the destruction of the parasite. There is thus in the alternate host a struggle for dominance between host and parasite, whereas under normal, favourable circumstances in the natural host there is mutual tolerance. Such is the general rule; nevertheless, many instances can be quoted in which parasites have little host specificity, as for instance *Trichinella* and *Toxoplasma*, and in which diseases such as psittacosis can be transmitted between groups having no ecological or familial relationships. Nevertheless, the most surprising anomalies occur over susceptibility to pathogens, for which no rational explanation has hitherto been found. Perhaps the greatest dangers to man arising from his role as host to scavengers come from contamination of his food supplies by organisms of the *Salmonella* group. *Salmonellae* are harmless commensals found in reptiles, particularly snakes, lizards and tortoises; the excreta of house lizards in particular can easily contaminate food supplies in hot countries. *Salmonellae* are also frequently carried by both birds and rodents, both of which are commonly responsible for contamination of food supplies.

Potentially, his simian cousins are the most dangerous animals to man. These simian pathogens have diverged insufficiently to be rejected by man, but far enough to provoke dangerous reactions in him. The familial aspects of primate zoonoses will be studied in the next chapter and it will be shown that, although the various members of the primate group diverged from the ancestral stock at some distance in point of time, yet phylogenetically they have differed little. At a time of climatic change, some sixty million years ago in the Oligocene epoch, ancestral primate forms radiated into a new habitat, comprising a thick belt of tropical forest which girdled the earth's belly. Climatic oscillations between ice-ages and interglacials, having remote effects on the tropics, resulted in great variations in the extent of this arboreal habitat, and some primate groups became partially or wholly, terrestrial. Nevertheless, man apart, the primates have remained wedded to the areas which they initially colonised and they still live mostly in the trees. This means that they have become adapted to a specialised type of life, specialised feeding habits, a specialised means of locomotion, in addition to which they have developed disease cycles shared with a different fauna from that associated with the terrestrial animals. So little is known of these cycles that no attempt can be made in this work to describe them. It can only be said that probability points to their involving other arboreal creatures, such as birds, bats and invertebrate fauna, which inhabit the upper arcades of the forests. Thus, primates are ecologically separated from terrestrial creatures, including man, and students of zoonoses must take account not of the contacts, direct or indirect, between simians and man but between the simian disease cycles and those in which man is involved.

This point may be illustrated by the instance, often quoted, of yellow-fever. The virus of yellow-fever is a natural and relatively harmless pathogen in monkeys of the genus *Cercopithecus* living in the upper arcades of tropical forest in Central Africa. It is transmitted by mosquitoes, *Aëdes africanus*, which are also parasitic on the monkeys in the upper arcades and which live and breed at treetop level, rarely descending to the ground. There is ecological separation of the vectors from the human species, and for this reason humans only rarely become infected. When, however, a human being infected in the forest goes to the town, he can easily start an epidemic of this very fatal disease, which is spread by the mosquitoes, *A. aegypti*, parasitic on man. In this way, yellow-fever spread from Central tropical Africa to the west coast of Africa and thence to Central America, where a new endemic focus was started among monkeys living in the Amazon jungles. The disease, on the face of it, gives an instance of the transmission (zoonosis) of a virus from wild monkeys to man, and from man to other distantly related groups of wild monkeys in another hemisphere (anthroponosis). It must, however, be noted that in this case the monkey is not the reservoir of yellow-fever virus. This is a mosquito virus; monkeys suffer transitorily from infection and are immune and free from the virus after recovery. The monkey in this cycle becomes infected when bitten by a mosquito and infects another mosquito when attacked at the height of viraemia; it is thus an intermediate host in a mosquito disease. In the cases of both yellow-fever and other arbo-viruses mammalian hosts other than monkeys may be involved, and such is indicated by the results of extensive serological tests. It is to be noted too that, although harmless to monkeys in Central Africa, yellow fever causes many deaths amongst the more unaccustomed South American monkeys.

This example illustrates that zoonosis between simian primates and man in feral conditions involves contact between two disease cycles, normally separated from each other. This contact requires vertical transmission from the forest canopy to ground level, and it may occur in a number of different ways. It may happen if the monkey itself descends to the ground; it may occur if arthropod vectors, either because of habit, or accidentally, descend to the ground and prey on terrestrial species; it may happen via mammals other than the simian primates involved in the disease cycle; or perhaps through the agency of creatures such as bats or birds which frequent both arboreal and terrestrial habitats. In South America, the vampire bats *Desmodus* spp. commonly attack both monkeys and cattle and occasionally man, and are known to transmit diseases such as rabies, and various blood protozoa (Hoare, 1965).

It would appear that the opportunities for transmission between arboreal and terrestrial disease cycles are fairly extensive. There is good ground for believing, nevertheless, that exchanges are rather rare, and it may well be that some unexplained anomalies, such as when diseases have appeared inexplicably and unexpectedly in certain areas, can be ascribed to an intrusion of pathogens from the arboreal cycles. Such instances will be studied in the ensuing text; meanwhile a further word of warning may be added. Almost all pathogenic organisms, or potentially pathogenic organisms, may suffer mutations affecting their virulence when passed from natural to unnatural hosts. With viruses in particular, virulence and invasiveness may be enormously altered, so that a relatively harmless virus, after passing through an unharmed animal, may become a deadly killer; conversely, its virulence may be reduced. Enhanced virulence could well arise from the transfer of a virus from an arboreal disease cycle into an unaccustomed terrestrial one, the end point being wholesale massacre of the human race in contact areas.

2 The Affinities of the Groups Comprising the Order Primates

Ivan T. Sanderson (1957) felicitously adopted for his book on monkeys the title *The Monkey Kingdom.* This was a more than usually happy selection of title because the primates, perhaps more than any other group in the animal kingdom, colonised a habitat that was virtually unoccupied. This occurred at a time some sixty million years ago in the Oligocene epoch. They diverged from the main line of mammalian evolution at a very early stage and developed along lines dictated by the new home they found in the vast Eocene forests, which covered a very wide belt of at least 50° North to South on either side of the Equator. Their closest relatives are the most primitive existing eutherian mammals, the Insectivora; and the most primitive members of the primates, the tree-shrews, still have points of affinity with the insectivores. The attached table giving the living genera of the primates is taken from *A Handbook of Living Primates* by Napier (in press), and the map giving the distribution of living primates has also been supplied by him. Table 3, which shows the time scale of human evolution is taken from Carrington (1963).

A comprehensive survey was made of the suitability of different primate groups for research purposes at a symposium held at the London Zoo in 1965, and jointly sponsored by the Zoological Society of London and the World Health Organisation (Fiennes, 1966). At this symposium research workers were especially urged that primates they use be precisely identified, so that results of their researches should not be invalidated. Different groups of primates have very different susceptibilities to

Sub order	Infra order	Super family	Family	Sub-family	Genus
Prosimii	LEMURI-FORMES	Tupaioidea	Tupaiidae	Tupaiinnae	*Tupaia* *Anathana* *Urogale* *Dendrogale*
				Ptilocercinae	*Ptilocercus*
		Lemuroidea	Lemuridae	Lemurinae	*Lemur* *Hapalemur* *Lepilemur*
				Cheirogaleinae	*Cheirogaleus* *Microcebus* *Rhaner*
			Indriidae		*Indri* *Avahi* *Propithecus*
		Daubentoni-oidea	Daubentoni-idae		*Daubentonia*
	LORISI-FORMES		Lorisidae	Lorisinae	*Loris* *Nycticebus* *Arctocebus* *Perodicticus*
				Galaginae	*Galago* *Euoticus*
	TARSII-FORMES		Tarsiidae		*Tarsius* *Callithrix*
		Ceboidea	Callitrichidae	Callitrichinae	*Cebuella* *Saguinus* *Leontideus*
				Callimiconinae	*Callimico*
			Cebidae	Aotinae	*Aotus* *Callicebus*
				Pitheciinae	*Pithecia* *Chiropotes* *Cacajao*
				Alouattinae	*Alouatta*
				Cebinae	*Cebus* *Saimiri*
				Atelinae	*Ateles* *Brachyteles* *Lagothrix*

ame in Common Usage	Geographical Distribution
ee shrew	S.E. Asia from N. India, Burma and S.W. China to Malaya, Sumatra, Java and Borneo
dian tree-shrew	Southern India
ilippine tree-shrew	Mindanao, Philippines
ooth-tailed tree-shrew	Vietnam and Borneo
n-tailed tree-shrew	Malaya, Sumatra and Borneo
mur entle lemur ortive lemur	Madagascar
warf lemur ouse lemur rk-marked dwarf lemur	Madagascar
dris oolly lemur faka	Madagascar
ye-Aye	Madagascar
ender loris	Southern India and Ceylon
ow loris	S.E. Asia from Assam and Burma to Malaya, Sumatra, Java and Borneo
ngwantibo	Central Africa from E. Nigeria to R. Congo
tto	W. and Central Africa from Senegal to Uganda
ushbaby	Africa from south of the Sahara to the Orange River
eedle-clawed Bushbaby	Central Africa from R. Niger to Lake Albert
arsier	Sumatra, Borneo, Philippines and Celebes
larmoset	S. America: Brazil, south of R. Amazon; E. Bolivia
ygmy marmoset	S. America: N. Peru, S. Colombia, W. Brazil
amarin, Pinché	Central and S. America: Canal Zone, N.W. Colombia, Amazon basin
olden lion tamarin	S. America: E. Brazil
oeldi's marmoset	S. America: E. Brazil, E. Peru
ouroucouli	Central and S. America: Panama to Paraguay
iti	S. America: Amazon basin from R. Meta south to Paraguay; N.E. Brazil
aki	S. America: Venezuela, south of R. Orinoco; the Guianas, Amazon basin
earded saki	S. America: The Guianas, north and south banks of lower R. Amazon into the Mato Grosso
Jakari	S. America: W. Brazil, N.E. Peru
Iowler	Central and S. America: Southern Mexico to Paraguay and S.E. Brazil
apuchin	Central and S. America: Honduras to S.E. Brazil, Paraguay and N. Argentina
quirrel monkey	Central and S. America: Costa Rica and Panama; south and east of R. Orinoco; Amazon basin
pider-monkey	Central and S. America: Southern Mexico to Colombia; the Guianas, Amazon basin, N. Bolivia
Voolly spider-monkey	S. America: S.E. Brazil
Voolly monkey	S. America: Amazon basin, Colombia, Ecuador, Peru

A CLASSIFICATION C

Sub order	Infra order	Super family	Family	Sub-family	Genus
Anthropoidea		Cercopithecoidea	Cercopithecidea	Cercopithecinae	*Macaca*
					Cynopithecus
					Cercocebus
					Papio
					Mandrillus
					Theropithecus
					Cercopithecus
					Allenopithecus
					Erythrocebus
				Colobinae	*Presbytis*
					Pygathrix
					Rhinopithecus
					Simias
					Nasalis
					Colobus
		Hominoidea	Hylobatidae		*Hylobates*
					Symphalangus
			Pongidae		*Pongo*
					Pan
					Gorilla
			Hominidae		*Homo*

LIVING PRIMATES (*continued*)

ame in Common Usage	Geographical Distribution
acaque	N. Africa, Gibraltar: Asia, from E. Afghanistan to China and Japan, south to India and Malaya, Sumatra, Java, Borneo, Philippines and Celebes
elebes black ape	Celebes
angabey	Central Africa from Senegal to E. Kenya
aboon	Africa, from the Sahara to the Cape; from Sierra Leone to Somalia; into Asia (Aden and Yemen)
rill, mandrill	Central Africa from E. Nigeria to R. Congo
elada baboon	Northern Ethiopia
uenon	Africa from Senegal, the Sahara Desert and Ethiopia to S. Africa
llen's Swamp	Africa: Congo basin
atas Monkey	Africa: from Senegal, the Sahara and Sudan to north of the Congo basin; Kenya, Uganda
angur	S.E. Asia: from India, S. Tibet, Burma and S. China to Malaya, Sumatra, Java and Borneo
ouc langur	Laos, Vietnam, Hainan
nub-nosed langur	W. China, N. Vietnam
agi Island langur	Mentawi Islands (W. coast of Sumatra)
roboscis monkey	Borneo
uereza	Africa from Gambia to Ethiopia and south to Angola, Tanganyika and Malawi
ibbon	S.E. Asia: Assam, Burma, S.W. China, Vietnam south to Malaya, Sumatra, Java, Borneo and the Mentawi Islands
iamang	Malaya and Sumatra
rang-utan	Sumatra and Borneo
himpanzee	West and Central Africa: from Gambia to Uganda; Congo basin from W. coast to Lake Victoria
orilla	Central Africa: Cross river south-east to L. Congo; Bondo district, Uele Valley; S.W. Uganda and E. Congo.
an	

DISTRIBUTION OF NON-HUMAN PRIMATE FAMILIES

1. TUPAIIDAE This family consists of the tree-shrews which comprise four genera: *Tupaia*, *Dendrogale*, *Urogale* and *Anathana*. Geographically they are widely distributed throughout the Far East.

2. LEMURIDAE A Madagascan family consisting of *Lemur*, *Hapalemur*, *Lepilemur*, *Cheirogaleus*, *Microcebus* and *Phaner*.

3. INDRIIDAE This family includes the long-legged Madagascan genera *Indri*, *Propithecus* and *Avahi*.

4. DAUBENTONIIDAE A monotypic family containing the aberrant Madagascan genus *Daubentonia*.

4. LORISIDAE A family which has representatives in the Far East (*Nycticebus*, *Loris*) and in Africa (*Perodicticus*, *Arctocebus*, *Galago*, *Euoticus* and *Galagoides*).

6. TARSIIDAE A monotypic family containing the single living representative, *Tarsius*, which is found on certain islands in the East Indies.

7. CALLITRICHIDAE Marmosets and tamarins. South American family consisting of five genera: *Callithrix*, *Cebuella*, *Saguinus*, *Leontideus* and *Callimico*.

8. CEBIDAE The largest family of South American monkeys containing all genera except the marmosets and tamarins, e.g. *Aotus*, *Callicebus*, *Pithecia*, *Chiropotes*, *Cacajao*, *Cebus*, *Saimiri*, *Alouatta*, *Ateles*, *Lagothrix* and *Brachyteles*.

9. CERCOPITHECIDAE Monkeys of the Old World (Africa and Asia). An historically continuous family now subdivided geographically into *Presbytis*, *Nasalis*, *Simias*, *Rhinopithecus*, *Pygathrix*, *Cynopithecus* (Asia); and *Papio*, *Colobus*, *Cercopithecus*, *Erythrocebus*, *Theropithecus*, *Miopithecus*, *Cercocebus*, *Mandrillus* (Africa). Only one genus, *Macaca*, is common to both regions.

10. HYLOBATIDAE Gibbons. Lesser apes of the Far East ranging from Assam in the west to Borneo in the east. The family includes two genera: *Hylobates* and *Symphalangus*.

11. PONGIDAE Great apes of the Old World. Africa: *Pan* and *Gorilla*. South-east Asia: *Pongo*.

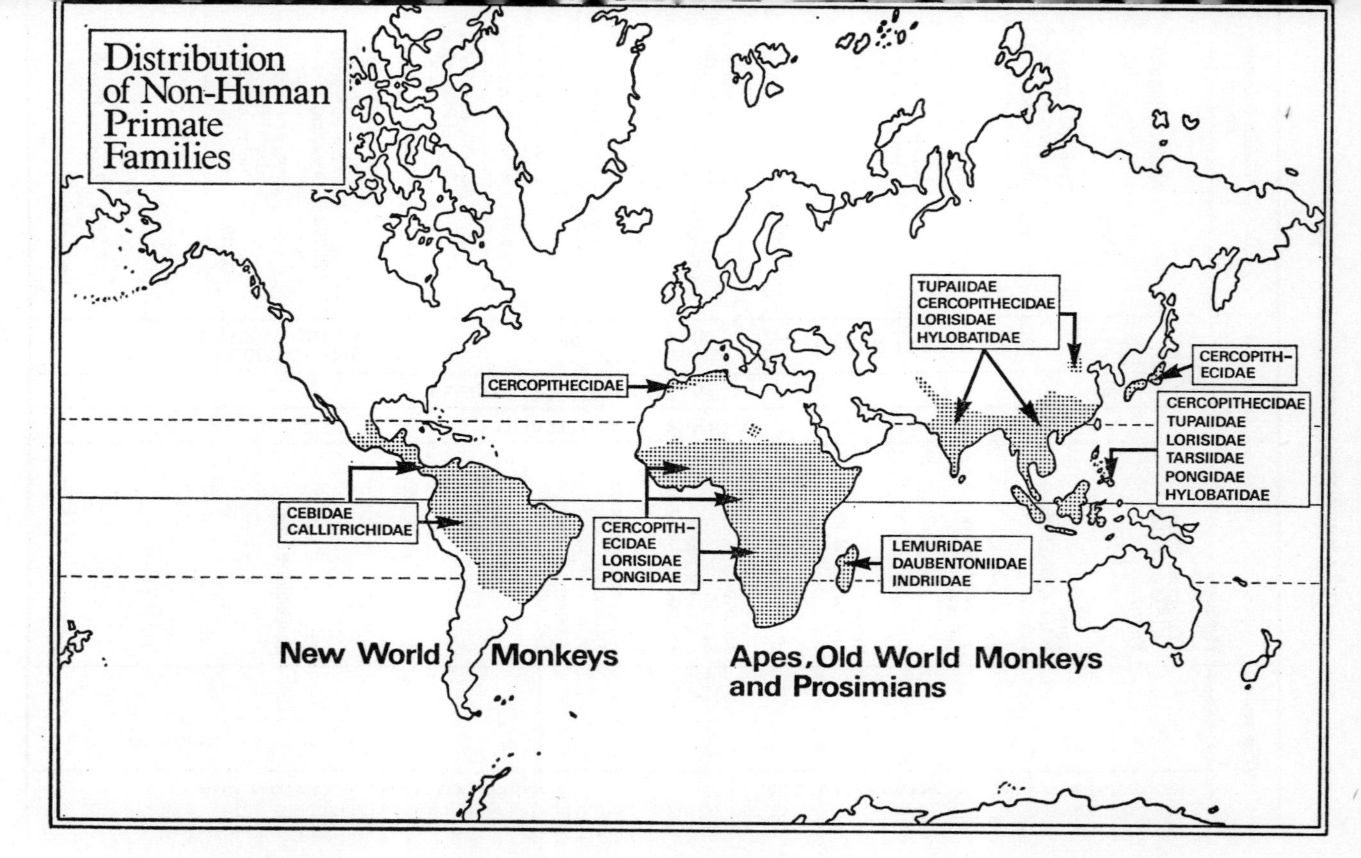
Distribution of Non-Human Primate Families
CERCOPITHECIDAE
TUPAIIDAE
CERCOPITHECIDAE
LORISIDAE
HYLOBATIDAE
CERCOPITH-
ECIDAE
CERCOPITHECIDAE
TUPAIIDAE
LORISIDAE
TARSIIDAE
PONGIDAE
HYLOBATIDAE
CEBIDAE
CALLITRICHIDAE
CERCOPITH-
ECIDAE
LORISIDAE
PONGIDAE
LEMURIDAE
DAUBENTONIIDAE
INDRIIDAE
New World Monkeys
Apes, Old World Monkeys
and Prosimians

CHART OF QUATERNARY TIME

Time scale	
	Holocene (10,000→)
UPPER PLEISTOCENE	Transition Phase (15,000–10,000) Fourth Glacial Phase 80,000 Third Interglacial Phase 190,000
MIDDLE PLEISTOCENE	Third Glacial Phase 240,000 Second Interglacial Phase 440,000
LOWER PLEISTOCENE, OR VILLAFRANCHIAN	Second Glacial Phase 480,000 First Interglacial Phase 550,000 First Glacial Phase 600,000 Preglacial Phase 1,000,000

Geological Divisions			Years Ago	Main Climatic Phases
QUATERNARY PERIOD	HOLOCENE OR RECENT EPOCH		10,000	Modern climates
	PLEISTOCENE EPOCH	UPPER PLEISTOCENE	15,000	Transition phase; gradual retreat of the ice-sheets
			80,000	Fourth, or Würm, Glacial Phase (Wisconsin Phase in North America)
		MIDDLE PLEISTOCENE	190,000	Third Interglacial Phase (Sangamon Phase in North America)
			240,000	Third, or Riss, Glacial Phase (Illinoian Phase in North America)
			440,000	Second, or Great, Interglacial Phase (Yarmouth Phase in North America
		LOWER PLEISTOCENE OR VILLAFRANCHIAN	480,000	Second, or Mindel, Glacial Phase (Kansan Phase in North America)
			550,000	First Interglacial Phase (Aftonian Phase in North America)
			600,000	First, or Gunz, Glacial Phase (Nebraskan Phase) in North America)
			1,000,000	Preglacial Phase; transition from Pliocene Epoch

CHART OF QUATERNARY TIME

Characteristic European Fauna	*Archeological Divisions*	*Main Cultures*	*Main Human and Sub-human Types*	*Years Ago*
	Modern urban society			
sting species st domestic animals	AGE OF METALS NEOLITHIC	Metal cultures and civilisation Advanced stone cultures	Modern races of *Homo sapiens*	
ɔpe fauna NDEER AGE (tundra ɹna)				
E OF WOOLLY AMMOTH olly mammoth *Iammuthus primi-nius)*	MESOLITHIC	Campignian Ertebӧllian Maglemosian Tardenoisian Azilian		10,000
olly rhinoceros *:hinoceros tichorhinus)* ve bear (*Ursus elaeus)* ve lion (*Panthera leo elaea)*	UPPER PALAEO-LITHIC	Magdalenian Solutrean Gravettian Aurignacian Chatelperronian	Fossil races of *Homo sapiens:* Cro-Magnon Grimaldi and Chancelade	40,000
E OF THE ANCIENT LEPHANT cient elephant *laeoloxodon antiquus)*	MIDDLE PALAEO-LITHIC	Mousterian	Neanderthal Man (*Homo neander-thalensis)*	160,000
mmuthus trogontheril rck's rhinoceros *:hinoceros mercki)* *ippopotamus major* *Vote:* During this per- the temperate and tropical climatic belts h their associated nas moved frequently th and south with the ance or retreat of the sheets. The above are r characteristic mam-ls of the time	LOWER PALAEO-LITHIC	Levalloisian Acheulian Clactonian Abbevillian Choukoutienian Oldowan	Heidelberg Man (*Homo heidelber-gensis)* Java Man and Pekin Man (*Pithecanthropus)* Australopithecinae and *Telanthropus* ? *Zinjanthropus* (exact status and date of this fossil is still speculative)	600,000
E OF THE SOUTHERN LEPHANT ıthern elephant *:lephas meridionalis)* uscan rhinoceros *:hinoceros etruscus)* re-toothed cat *Iachairodus)*	? Origin of tool-making tradition		? Pre-Australo-pithecine sub-hominids (no known fossil evidence)	1,000,000

This chart has been devised to give a general picture of the geological, climatic and archaeo- ical phases of the Quaternary Period, with their associated fossils and human cultures. Most the dates are still speculative and recent research suggests a rather greater antiquity for the stralopithecines than is shown here. The time scale on the left gives an approximate indication the duration of the various phases; the remainder of the chart is diagrammatic only and is not wn to scale.

different types of disease, and indeed when newly imported, often carry quite different types of parasites, depending on their origin. An attempt will be made here in the appropriate sections to indicate which parasites are likely to be naturally carried by which groups of primates and to state what is known about the general susceptibility of the different groups to different diseases. Too many people, when using primates, regard them simply as laboratory animals, without realising the tremendous variation between the different groups and even between individuals of one group. The more advanced groups of simian primates are intelligent, sensitive creatures and should be treated with consideration. Often their confidence can be won, as with dogs, and then research procedures can be more readily performed with their co-operation. It is particularly to be hoped that, in the course of time, minimum standards will be laid down for their cage sizes, thus eliminating much distress they suffer, and indeed disease, as a result of being confined in cramped quarters.

The most primitive living primates are the tupaias, or tree-shrews, small squirrel-like mammals found only in India, South-east Asia and Indonesia. These little animals have existed almost unchanged for fifty to sixty million years and combine characters found both in the primates and insectivores. Coon (1963) in his *Origin of Races* describes their character as follows: 'Their excesses of courage, wrath and libido constitute a caricature of uninhibited human behaviour, unparalleled by any larger and fully accredited primate.'

Co-existing with the tree-shrews in the Oligocene epoch, and undoubtedly derived from a common ancestor, were lowly 'pro-simian' forms of lemurs and tarsioids. The pro-simians, consisting of the lemurs, lorises, galagoes and tarsiers, are thus a group of extreme antiquity, still represented by many existing forms, although they appeared on earth between fifty and sixty million years ago. It is certain that the Old-World and New-World monkeys, Ceboidea and Cercopithecoidea, developed from them independently. The ancestors of the two great groups of the monkey kingdom, in spite of their many resemblances, thus diverged at a very early stage of primate evolution. It seems probable that the ape (Hominoidea) and Old-World monkey stocks (Cercopithecoidea) diverged almost as early; one well-known anatomist and anthropologist, Professor Wood Jones, believes that the human stock originated directly from the tarsioids and is thus no more closely related to apes and monkeys than to the pro-simians, but this seems unlikely. Anatomical considerations group the Hominoidea with the Cercopithecoidea, but they certainly diverged during the Oligocene epoch at least forty million years ago. It is significant that both man and the great apes share a similar blood-group system, and that their parasites in some instances belong to similar genera.

Much has been written in favour of racial affinities between man and the great apes, based on their sharing common ectoparasites. The position was reviewed by Zuckerman (1933) in his *Functional Affinities of Men, Monkeys and Apes*, and he also reviewed the relevant literature, which will be noticed in a later chapter. The genus of sucking lice, known as *Pediculus*, occurs on man in two forms, the head and the body-louse; other species of the same genus are found on all the anthropoid apes, but not on the Cercopithecoidea. Human lice can infest the anthropoids, and the human pubic louse, *Phthirus*, can infect gorillas, and has been found on them in the wild. Lice of a different genus, *Pedicinus*, infest all Cercopithecoidea but do not touch either the Anthropoidea or the New-World monkey species. On the other hand a species of *Pediculus*, (*P. pseudohumanus*), infests New-World monkeys, and this could indicate a relationship of the pro-simian forms from which Anthropoidea and Ceboidea were derived, closer than that with the ancestral forms of the cercopithecoid monkeys. It has been suggested that the New-World monkeys acquired their lice from American Indians, when their ancestors invaded the American continent. This hardly invalidates the argument, since the New-World monkeys are at least susceptible to *Pediculus*, whereas the Old-World monkeys are not.

There can in any case be no doubt whatever about the antiquity of the period when the various primate groups diverged from each other. The New-World monkeys undoubtedly arose by independent evolution from some ancestral pro-simian group. The Cercopithecoidea and Hominoidea too may have arisen independently from different pro-simian groups. The group that gave rise to Hominoidea possibly had affinities with that from which the ceboids developed, but it had the tooth formula of the cercopithecoids. The human stock also certainly diverged at a very early date from that which produced the anthropoid apes, probably in the Miocene. It has been suggested that the chimpanzees were derived independently from a protohuman stock, and that they reverted to an arboreal existence, which the protohuman stock had deserted. Some support for this view comes from Hamerton and Klinger (1963), who suggest, from chromosome studies, that the chimpanzees are more closely related to man than to the other anthropoid apes.

In spite of their wide distribution throughout Africa and Asia, the Cercopithecoidea represent a reasonably homogeneous group of monkeys. As with other groups of animals in a new habitat, they have radiated to occupy complementary 'niches'; some species inhabit the upper canopies of the forest; some the middle canopies; some the lower canopies; and some have become savannah species and spend part of their time on the ground. The baboons, patas monkeys and the macaques have tended to become secondarily ground-living, and others such as the talapoins have invaded specialised habitats such as swamp forests. The descent to the ground, as shown by the baboons, has necessitated means of protection from predators and has resulted in the development of the canine teeth as weapons and the adoption of a degree of social organisation. These adaptations parallel those which occurred in pre-human history, which also gave rise to the qualities of organisation and intelligence found only in *Homo sapiens*.

It has been suggested that the cercopithecoids were originally a stronger race and were able to secure for themselves the larger and stronger branches of the trees, as they do today. The weaker forms, which resembled the ancestral gibbons, were driven to the slenderer and less secure branches. Life in this habitat favoured the development of certain characteristics. Thus, monkeys progress through the forest by leaping in a four-footed manner from one thick branch to another, whereas anthropoids proceed by swinging on their arms from the smaller branches by a form of locomotion known as 'brachiation'. They thus entrust themselves to the branches for only short periods, quickly swinging from one to another, so that their weight can be supported. Amongst the heavier species like the gorillas, chimpanzees, and orang-utans, the feet developed prehensile properties so that the weight could be taken at four points instead of one or two. The primitive gibbons, as gibbons do today, must have run along branches on the soles of their feet gripping other branches above them with their hands, and thus moving at times in an erect manner. Man himself must have been derived from an anthropoid form without specialised foot structure, and this fact supports the theory of early separation.

Many workers consider that studies of the natural parasites of animal groups can give important clues to phylogeny. The position in relation to ecto-parasites has already been discussed. Man may share his ecto-parasites with the anthropoid apes, but the position is entirely different in relation to platyhelminth infestations. The larval stages of the human tapeworms, *Taenia saginata* and *T. solium*, are found in the muscles of the ox and the pig. This is an ecological association derived from a predator/-prey relationship over many millennia. A similar relationship has not been developed between such animals and the anthropoid apes or monkeys, since they have never become predators in the true sense of the word. On the other hand the cysticercal stages of various tapeworms are commonly found in wild apes and monkeys. The coenurus larvae of *Multiceps multiceps* or *M. serialis* are commonly found in the brains of baboons. Hydatids of *Echinococcus granulosus* have been found in lemurs, monkeys and baboons. Cysticercoid forms, the larvae of *Taenia* spp., have been found in the peritoneal cavities of gibbons.

The tapeworm most commonly found in adult form in simians is a tapeworm of herbivores belonging to the genus *Bertiella* of the Order Anoplocephalidae; it can infect man, but is rarely found. It is probably transmitted by mites. Whereas man is an undoubted predator, simian primates are potential prey species, a distinction of some importance in their parasitology.

Apes and monkeys appear to be to a great extent resistant to the ascarid worms of man, but when they do become infected the disease caused is serious, and death may result. The human pinworm, *Enterobius vermicularis*, has been found in captive apes, gibbons and marmosets, and it is evident that cross-infection from man to monkey or ape can occur. The commonest nematode infections of monkeys and apes are due to species of *Oesophagostomum*, a genus which is very rare in human beings. Amongst the Protozoa we find instances, as with amoeba and balantidium, where apes carry parasites identical to those of man; with the plasmodia some species are no longer cross-infectious, whereas others, although morphologically different, can pass from man to ape and vice versa. These matters are discussed in detail in succeeding chapters, together with the evidence in relation to bacteria and viruses. With some parasites the divergence in ways of life between human and simian primates has led to new host/parasite relationships, whereas in others a continuous relationship has persisted in both groups in such a way that the parasite may be virtually unaltered, although it may have changed in virulence, infectivity and other respects.

The ecological and familial factors in primate zoonoses are therefore confused. The various groups in the one order range from a very primitive mammal to *Homo sapiens*, anatomically unspecialised, but in respect of his social evolution the most uniquely specialised animal the world has yet seen. The divergence between the groups occurred in a very remote era, in spite of which anatomically and physiologically they have retained many features in common. With regard to the parasites present in the various groups, there has in some cases been divergence of such extent that cross-infection can no longer take place. In some cases the parasites appear to have diverged little; in other cases host tolerance and resistance have been lost, and the results of cross-infection can be serious. In succeeding chapters the different groups of parasites will be studied, and an attempt will be made to discover to what extent the parasites of one group can be dangerous to the other and particularly to man. The inclusion of a chapter on the diseases of wild-living simians was intended, but information is so scanty as to make the task a fruitless one. References to such diseases are quoted by Halloran (1955).

Part II Parasites

3 Ectoparasites

There is, unfortunately, little reliable knowledge about the extent to which the various groups of primates suffer from ectoparasites under wild conditions, or indeed what are the main parasites which attack them. It is, however, certain that during their adaptation to an arboreal habitat certain parasitic forms have become adapted to them and these parasites are somewhat specialised. Ectoparasites as such are probably of relatively small importance, although the arthropod vectors – studied in succeeding chapters – are of great significance particularly in relation to the primate zoonoses.

In both the wild state and in captivity, apes and monkeys have reasonably clean skins and appear able, as a result of their social habits, to maintain them in good condition and relatively free from parasites. This is especially the case when monkeys are kept together, less so when in isolation. All groups of primates endlessly 'groom' each other, picking off scurf and scabs and, when present, parasites and their eggs. The parasites are usually placed in the mouth and eaten, a habit which is effective in disposing of them but – as will be seen – a potential source of danger. Lung mites, with which most rhesus monkeys are affected, may possibly gain access to the body in this way.

Grooming itself is not an indication that parasites are present. It is a social activity with obvious hygienic merits, which has become part of the behaviour pattern of the group, of significance in the relationships between individuals and in the social hierarchy system. Scratching does not necessarily indicate the presence of ectoparasites or even of skin irritation. In many monkeys, it is a sign of embarrassment or diffidence. Nevertheless, skin lesions are often difficult to detect on monkeys because of their thick hair, and in cases of persistent scratching the skin requires careful examination to determine whether lesions or parasites are present.

Acari (Mites)

Trombidiidae In the Pacific region of Asia, monkeys are frequently found infested with skin mites (Trombiculidae), and these have also been described from monkeys by Trapido and Work (1962) in India. Such mites belong to the group which contains the harvest bugs, or harvest mites, and are of importance in the Far East as transmitters of scrub-typhus and scrub-itch. Primates are thus potentially dangerous to man as hosts of these, and human cases infected by monkeys are recorded by Blanc and Woodward (1945) and Womersley (1952)

Sarcoptidae (Mange Mites) The first description of sarcoptic mange in simians was that of Fuerstenberg (1861), who regarded monkeys as definitive hosts of *Sarcoptes scabiei*, and possibly also of another species, *S. mutans*. Pillers (1921), however, regarded scabies as a rare disease, although he described a case in a two-year-old chimpanzee. This is apparently the only record of the infection in a chimpanzee, though Philippe (1948) described a case of *Demodex canis* infection in a chimpanzee which had been contracted from a dog with which it played.

In the orang-utan, scabies was described by Weidman (1923); and Rewell (1948) successfully treated a case in a gorilla with benzyl benzoate emulsion.

In gibbons and siamangs, an outbreak of scabies in which nine of the ten affected animals died, is described by Ruch. Nine persons who were in contact with the gibbons acquired the infection from them, but the disease in humans was mild. Ratcliffe (1955) also described scabies in gibbons complicated by a fungal infection. The scabies was successfully treated, and the fungus disappeared naturally when the gibbons were moved to a large outdoor cage.

There is, therefore, sufficient evidence to show that the higher primates can act as hosts to *S. scabiei* or related species. The infection is especially severe in gibbons, and is transmissible from them to man but in a less severe form.

Amongst the lower primates, scabies appears to be a rare condition; possibly because it is so severe in its effects that few infected animals survive. Ruch quotes Brug and Haga (1930) and Delorme (1926), who infected monkeys experimentally. Philippe described an outbreak in the former French Guinea (Republic of Guinea), which affected *Cercopithecus aethiops* (grivet monkey) and *Papio papio* (Guinea baboon). The grivets all died as a result of the infection, but the baboons responded to treatment, although they became affected by a drug-induced dermatitis. The author considered the parasite to be a distinct species, which he named *S. pitheci.* This claim, if established, would suggest that sarcoptic mange is an independent, if rare, disease of wild-living monkeys in Africa. The only other record appears to be that of Fox (1926) who described scabies in a drill (*Mandrillus leucophaeus*).

There appear to be no records of scabies in pro-simians or in ceboid species.

Pneumonyssidae (Pulmonary Acariasis) In the opinion of many workers, including Innes *et al.* (1954), pulmonary acariasis is present in the great majority, if not all, of wild-bred Old-World macaques. It has also been described by Lapin (1956) as occurring in colony-bred monkeys at the Primate Centre at Sukhumi, Georgia, USSR. It is, however, by no means confined to the Old-World monkeys and has been described in *Lemur mongoz* (*Liponyssus madagascariensis*) by Hirst (1921), and by Vitzthum (1930) in *Lagothrix infumatus* (the smoky woolly monkey of South America). The wide incidence of lung mites in monkeys may be attributed to habits of grooming, since transmission is believed to occur by ingestion during the course of grooming. Some workers, however, believe that infection is by way of inhalation.

It has been suggested by Davis (1945) that transmission of these mites from monkeys to man is responsible for certain pulmonary disorders of man described by Carter *et al.* (1944) and by Soysa and Jayarwadena (1945). These include asthma, bronchitis and tropical eosinophilia of the lung (Loeffler's syndrome).

Hamerton (1939) attributed a number of deaths at the London Zoological Gardens to mite infestation; Helwig (1925) also considered these mites to be the cause of deaths; and Grzimek (1951) described an acute case in a chimpanzee of fourteen years; nevertheless clinical signs are generally rare and not severe. In a personal communication, Innes (1965) summarises his findings as follows:

> It is safe to say that no imported *Macaca mulatta* will be found without some degree of mite infestation and lesions. By casual inspection of living monkeys, clinical signs of pulmonary disease may not be detectable, or may be only those of moderate attacks of sneezing and coughing. It is impossible to decide clinically whether any given living monkey will show no, a few, or disseminated pulmonary lesions at necropsy. The assumption must be that all imported, adult *Macaca mulatta* are infected. It seems inescapable that mites, by their wanderings in the lungs, must irritate the bronchial passages and interfere with normal health, and reduce pulmonary capacity. Routine roentgenograms of the chests of monkeys in a normal colony, or by importers, however, seem of no diagnostic value. Since the pulmonary lesions are parasitic in origin, a prominent participation of eosinophilic leucocytes in the histologic reaction was suggestive that this might be reflected in an eosinophilia of the peripheral blood. This is not always evident. Complete routine haematologic examinations have been made, for other reasons, on very many monkeys passing through our hands. Any marked degree of eosinophilia, particularly that which can indubitably be correlated with the presence of mite lesions, is not constant.
>
> Pulmonary lesions and the mites have been found in monkeys which have died within a few days of arrival, and in some killed on receipt within as little as seven days after leaving India. It is certain that the animals acquire this infestation in their native country, India, Burma, or elsewhere in the Far East. Beyond that, there is little or no information regarding the life history of the mite and the pathogenesis of the disease.

The pathological and histological findings of this condition have been described in great detail by Innes *et al.* 1954.

The parasite chiefly concerned is *Pneumonyssus simicola* (Banks, 1901) but a number of different species have been described. The validity of these has been questioned; Fuhrman (1954) reviewed the genus and recognised *P. simicola*, *P. congoensis*, *P. duttoni*, *P. dinolti* and *P. stammeri*, as valid on morphological grounds. Subsequently Boehm and Supperer (1955) established further species in *P. oudemansi*, recovered from a chimpanzee, and *P. vitzthumi* from an orang-utan.

The importance of this condition as a zoonosis has yet to be established.

Linguatulidae Heuschele (1961) has recorded internal parasitism of a Hamlyn's owl-faced monkey (*Cercopithecus hamlyni*) and a talapoin (*Miopithecus talapoin*) with nymphs of *Armillifer armillatus*. The parasites were encysted in the mesentery, omentum, liver and spleen. They are also common in the mesentery of baboons from various parts of Africa (Wright, 1965). Infestations are also reported by Ruch as occurring in the liver, and he quotes a report by Desportes and Roth (1943) of infestation in a capuchin monkey kept in a European zoo. Both larval and nymphal stages are common in man (Fain, 1960), (Watson, 1960). The nymphs cause no untoward symptoms. Both monkeys and man are cast for the role of intermediate hosts, the adults infect snakes, especially pythons. A long list of primate hosts for this parasite is given by Stiles and Hassall (1929), including man, chimpanzee, baboon, macaque and others.

Related parasites, *Porocephalus cercopitheci* and *P. clavatus* are also reported from both *Homo sapiens* and African primates, including *Cercopithecus* sp. and baboons. The larvae are found in the mammalian hosts, which include pigs; the adults in the respiratory passages of snakes.

Linguatula serrata, the tongue worm of the dog, has also been reported in both man and simians. Either larvae or adults may be parasitic.

Insecta (Fleas, Flies, Lice)

Siphonaptera (Fleas) There is little information on the extent to which simians are infested with fleas under wild conditions. No cases in captive monkeys are quoted by Ruch, and such information as is available has been listed by Stiles and Hassall: *Tunga penetrans* (Pulex) in the gorilla; *Echidnophaga gallinacea* in the grey lemur (*Myoxicebus griseus*); *Ctenocephalus canis* in *Colobus* spp.; and *Ctenocephalus felis* in a slow loris (*Nycticebus* sp).

These few records of fleas, mostly natural parasites of other groups of animals, give no indication of the importance of fleas in simians, or of any possible role in disease transmission to man.

Dermatophilus (*Tunga*) *penetrans*, the jigger flea, was introduced into the Old World from South America; it penetrates beneath the skin causing great irritation. Its presence in a gorilla is interesting, but of little significance since it has little host specificity, and domestic animals are the normal reservoir for infection of man. Haddow (1964) states that it can cause considerable trouble in captive monkeys, particularly in guenons and baboons. This occurs when it enters and becomes firmly lodged in the hard skin pads on the buttocks. There is risk of infection after removal from such a site.

Diptera (*Cuterebra*, *Cordylobia*) Infection of various species of howler monkey (*Alouatta* spp.) with larvae of oestrid flies of the family Cuterebrinae is described by Shannon and Greene (1926). Such parasitism in monkeys and man has been described by a number of expeditions to Central and South America, including that of Fawcett (1953). Spider monkeys (*Ateles*) are apparently unaffected. The bots are often very common under the skin of the neck.

A rather similar fly, *Cordylobia anthropophaga*, is found in East and Central Africa, but a wider range of host is tolerated by the larvae. Man, monkeys and both wild and domestic animals such as dogs may be affected.

Mallophaga (Biting Lice) The biting lice are relatively rare on simians, and appear to have no significance in relation to zoonoses. Stiles and Hassall list six species, which are nevertheless of some interest as showing the world-wide distribution of related species amongst the simian genera. *Trichodectes armatus* is reported from Brazil as parasitising woolly spider monkeys (*Brachyteles*). *T. colobi* was recovered in the Kilimanjaro area of

East Africa from *Colobus* sp. *T. mjoebergi* is described from North Borneo as a parasite of the loris, *Nycticebus* sp. From South America, *T. semiarmatus* is recorded from the howler monkey, *Aloutta* sp. Two unidentified species of *Trichodectes* are also recorded from howler monkeys. A biting louse, *Trichophilopterus babakotophilus*, is recorded as a parasite on the lemur *Indri indri* in Madagascar; and *Tetragynopus aotophilus* on the douroucoulis, (*Aotus boliviensis*), in Bolivia. Hopkins (1949) quotes *Trichophilopterus ferrisi* as a parasite common on the mongoose lemur (*Lemur mongoz*) in Madagascar.

In addition to the records quoted by Siles and Hassall, there is a single description by Fenstermacher and Jellison (1932) of infestation of *Macaca mulatta* with the porcupine-biting louse (*Eutrichophilus setosus*). These monkeys were kept near a cage of porcupines which were plainly the source of infection.

There are no records of Mallophaga as parasites of man or the anthropoid apes.

Anoplura (Sucking Lice) Hopkins (1949) in his *Host Associations of Lice of Mammals* gives the following lists of sucking lice which are parasitic on simians (see also Stiles and Hassall):

Tree shrews	*Anathana ellioti*	*Docophthirus acionetus*
Lemurs	*Propithecus diadema*	*Phthiropediculus propitheci*
Galago	*Galago senegalensis* and *G. demidovii*	*Lemurphthirus galagus*
Sakis	*Pithecia monachus*	*Pediculus lobatus pseudohumanus*
Uakaris	*P. rubicundus*	*P. l. pseudohumanus*
Howlers	*Alouatta caraya* *A. beelzebul* *A. villosa* *A. fusca* *A. seniculus*	*P. l. pseudohumanus*
Spider monkeys	*Ateles* spp.	*P. l. pseudohumanus* *P. l. atelophilus*
Marmosets and Tamarins	*Callithrix santremensis* *Leontideus nigricollis*	*Harrisonia uncinata* *Gliricola pintoi*
Gibbon and siamang	*Hylobates* spp.	*Pediculus humanus friedenthali*
Chimpanzee	*Pan satyrus* spp.	*P. schaeffi*, *Phthirus pubis*

Gorilla	*Gorilla* spp.	*Phthirus gorillae*
Man	*Homo sapiens*	*Pediculus humanus humanus*, *P. h. capitis* and *Phthirus pubis* (one record of *Pediculus pseudohumanus*)

All other primates, including macaques, cercopitheques, mangabeys, baboons, patas, langurs, proboscis and colobus, carry only *Pedicinus* spp.

In addition to the above, Cummings (1916) found *Pediculus humanus capitis* parasitising a red-faced spider monkey (*Ateles paniscus*); the infection appeared to have been acquired from a human source, showing that sucking lice can be shared by man and the New-World monkeys, but not the Old-World monkeys.

The sucking lice may thus be regarded as interchangeable between man and the anthropoid apes, and the New-World simians, with the possible exception of the marmosets and tamarins. Old-World monkeys are not affected by the sucking lice of man, anthropoids and New-World monkeys. There are, however, no records of transmission of rickettsial diseases by lice from anthropoids or New-World monkeys to man, or vice versa, though theoretically this would appear possible.

4 Arthropod Vectors

Of greater importance than the ectoparasites to the student of zoonoses are those external parasites, mostly arthropods, responsible for the transmission of infectious diseases caused chiefly by protozoa, helminths and viruses. A growing volume of knowledge, stemming originally from the researches of the scientists of the Rockefeller Institute on yellow fever, is beginning to illuminate the extent of our ignorance about reservoirs, vectors, parasites and infection cycles.

The parasites chiefly concerned belong to the ARBO (arthropod-borne) group of viruses, of which a large number, whose existence was unsuspected, have been discovered during the course of studies made on yellow fever. Some account of these will be given in a later chapter. Newer knowledge is demonstrating how the infection cycles of the parasites are intimately related to the bionomics of the vectors by which they are transmitted. A three-host tick may not be parasitic at all stages of its life history on the same animal species; it is thus able to convey a virus to quite dissimilar hosts, which may react differently to its presence. With mosquitoes too, it has been shown by Haddow and his colleagues in Uganda (Haddow, 1961a; 1961b) that forest mosquitoes may have a diurnal cycle, feeding for part of the day in the upper canopies of the forest, and at other times at ground level. The epidemiological implications are discussed at length by Haddow *et al.* (1964).

These varied habits of the vectors bring parasites into contact with a wide variety of species in some of which they may be 'silent' pathogens; and in others they may cause disease. Certain animals may be normal hosts of the vectors, but at times abnormal hosts may be attacked. Such an abnormal host may be man, and he may thereby be brought into contact with abnormal parasites with serious consequences. That such indirect contacts occur between the lower primates and man, and that diseases may pass between the two in this way, is well known from the work on yellow fever; newer evidence suggests that with malaria, too, cross-infection between monkeys and man may occur in some parts of the world, such as Malaya (Hoare, 1962).

Ixodoidea (Ticks)

On captive monkeys, the problem of ticks is unimportant since, when replete, the ticks fall off, and under captive conditions reinfection does not occur. Hoogstraal (1956) described the presence of *Rhipicephalus sanguineus* on adult captive grivets (*Cercopithecus aethiops*) in the Sudan, but supposed them to have originated from dogs. It is to be noted that this tick transmits canine tick-typhus, caused by *Rickettsia canis* or *R. conori*, which may also be transmitted to man, see Carmichael and Fiennes (1942). It is not known whether the rickettsia can also affect monkeys.

Hoogstraal found no ticks on Stuhlmann's guenons (*C. mitis stuhlmanni*), grivets (*C. aethiops*) or East African patas monkeys (*Erythrocebus patas pyrrhonotus*). Only one of over sixty bush-babies (*Galago* spp.) from two regions was infested, and that animal harboured only one tick. Amongst twelve colobus (*Colobus polykomos dodingae*) from a Sudanese forest, one harboured a single tick, but several from a group in Kenya were infested. Amongst baboons (*Papio* spp.) no animals from a group taken from one Sudanese province were infested. In another province, baboons from the communal bands were free from ticks, but a number of lone males were found to be infested, and one carried as many as two hundred ticks mostly in the axillary region.

Argasid ticks, such as *Ornithodorus*, important as transmitters of spirochaetal diseases such as relapsing fever, have only infrequently been described as parasites of simians but must surely be common in some regions and attack at night. Heavy tick infestations by *Haemaphysalis* spp. were, however, found by Trapido and Work (1962) during their work on Kyasanur Forest disease.

Stiles and Hassall list a number of tick species as being recovered from wild primates as follows:

Argas reflexus	from monkey
Ornithodorus talaje	*Macaca irus*
Ixodes calvipalpus	monkey and baboon (N.W. Rhodesia)
I. loricatus	*Ateles* sp. (Brazil, Mexico)
I. schillingeri	*Colobus caudatus* (E. Africa)
Haemaphysalis palmata	monkey (Cameroon and Sumatra)
Boophilus annulatus	*M. irus*

It is thus evident that wild primates are attacked by ticks in most areas in which they are found, and that both Old- and New-World monkeys are affected by a number of different species. The extent to which they are important as parasites of monkeys is not evident, nor is information given as to the stage of tick, larva, nymph or adult, which was collected. With regard to the relapsing-fever tick, there is no information as to the extent to which it preys on monkeys by emerging from the bark of trees, in which they rest for the night, although it is known that monkeys have favourite trees in which they sleep and it would be unlikely that these did not harbour the parasites.

Many simian species have been found infected in the wild with the spirochaetes of relapsing fever, including New-World species such as *Leontocebus*, *Cebus*, *Ateles* and *Aotus*, and Old-World species such as *Erythrocebus* and *Cercopithecus* spp. In particular Clark (1942) found naturally occurring infections in three specimens of *Leontocebus geoffroyi* and two *Cebus capucinus capucinus*. Infection also existed amongst marsupials (opossums), armadillos and grison. There thus existed the material for a complicated cycle of infection from tree to ground, in which man might well be involved. Clark found that the vectors were *Ornithodorus talaje* and *O. venezuelensis*; the former may select human, mammalian, avian or reptilian hosts, but the latter was regarded as the main vector for animal-to-man and for man-to-man infections. The infectious agent was *Spirochaeta neotropicalis*.

The infrequent records of ticks on monkeys is no doubt in part due to their grooming habits, and this is supported by Hoogstraal's observations that lone male baboons were heavily infected, whereas those in troops in the same area were free. Trapido (1964), however, states that the infestation of bonnet macaques (*Macaca radiata*) and langurs (*Presbytis entellus*) in Kyasanur Forest in South-west India was seasonal. Infestation was caused by a large number of species of *Haemaphysalis*, three-host ticks; during the dry season, the monkeys were infested with larvae and nymphs; the nymphs dropped from the hosts, when engorged, moulted on the ground and became adult; they then found their way to ground-living species, chiefly ruminants but including man. Thus, if the monkeys were examined for ticks during the wet season, it might be supposed that they were relatively free from infestation, although in the dry season they harboured large numbers. In this way Kyasanur Forest virus could be transmitted between monkey and man. Ruminants are important in providing a source of blood for adult *Haemaphysalis* of certain species (viz. *H. spinigera*, *H. turturis* and *H. bispinosa*); they do not appear to be important as reservoirs of Kyasanur Forest disease virus.

It is evident from Trapido's work that little reliance can be placed on records of tick infestation in simians – or for that matter in any other animals – where a two- or three-host tick is involved, unless observations were made throughout the different seasons of the year. It is probable that ticks will be found to be of greater significance in spreading disease from animals living in the upper arcades of the forest to those on the ground than has hitherto been supposed. Unlike the case in Africa, in India one need not look for a distinctive arboreal tick fauna as ectoparasites of monkeys, since the monkeys concerned, both *Macaca* and many *Presbytis*, particularly *P. entellus*, spend a great deal of time on the ground.

Culicidae (Mosquitoes)

The mosquitoes are of importance as vectors of viral, protozoal and helminth diseases, such as yellow fever, malaria and filariasis. Their importance in transmission of disease from one class of vertebrate to another depends to a great extent upon their feeding habits and way of life. In relation to tree-living primates, they have been studied to the greatest extent in the course of yellow fever and malaria research.

It has been supposed for many years that the different mosquito species are 'layered' in the forest arcades as are the simian species, and that those dwelling at high levels rarely, if ever, descend to the ground. It was thus supposed that contact, and so disease transmission, between tree- and ground-living species was at most sporadic and even accidental. Studies of the vertical stratification of forest mosquitoes in South America were described by Galindo *et al.* (1950) and by Trapido *et al.* (1955); these studies showed that the former view was only partially correct, since upper canopy mosquitoes could be caught at ground level under certain conditions of climate and forest cover. More recently Haddow and his colleagues (Haddow, 1961a,b, *et al.*, 1964) have studied the habits of East African mosquitoes in relation to stratification and have shown that some canopy species have a diurnal rhythm, feeding in the upper arcades during the daylight hours and descending to ground level to feed at dusk. Nevertheless, Downs *et al.* (1955) in Trinidad found no yellow-fever virus in the 118,000 ground-living mosquitoes which they caught, and none in the 120 canopy species caught at ground level (*Haemagogus*); pooled samples of *Haemagogus* caught at canopy level, on the other hand, contained yellow-fever virus.

It has for long been supposed that the contact between tree-living primates and man, in the transmission of yellow fever, depended rather on accidental circumstances than on natural contacts. This view is given substance by observations such as those of Bugher *et al.* (1944), who found that *Haemagogus capricornii*, responsible for the transmission of yellow fever in Eastern Columbia, were very rare at ground-level, but that they attacked wood-cutters viciously after a tree was felled. They had been obviously concealed in the foliage.

It may be accepted therefore that contact between mosquitoes, which prey on tree-living primates, and ground-living species including man, may at times be associated with vertical movements of the mosquitoes, but it is also the result of accidental circumstances such as when trees are felled in the forest and the tree-cutters are bitten. Both main groups of mosquitoes, *Culicini* and *Anophelini*, are represented by tree-living species. The former are associated with the transmission of yellow fever, dengue, probably many other virus diseases and a number of diseases caused by nematode worms of the order *Filarioidea*. Anopheline mosquitoes transmit malaria, and it has been suggested that a new problem has arisen in Malaya by which contacts between monkeys and man have been established through these mosquitoes; if this is so, monkey malarias may sometimes affect the human population and human malaria the arboreal primates, Hoare (1962). Anopheles mosquitoes, as well as the Culicines, also transmit filaria (Bertram *et al.*, 1958).

The species of mosquitoes concerned in yellow-fever transmission are well known. The original dispersal area of yellow fever was undoubtedly Central Africa, where virtually all species of arboreal primates become naturally infected, but do not suffer severe symptoms (Haddow *et al.*, 1951). Lumsden and Buxton (1951) failed to find ground-living species (baboons) with positive sera. Nevertheless, Haddow *et al.* found six of twenty positive baboon sera in Uganda, and Smithburn and Haddow (1946) isolated yellow fever virus – and Haddow and his colleagues Zika virus – from mosquitoes taken at ground level. The vectors in Africa are various species of *Aëdes*. *A. aegypti* is responsible for the transmission of urban yellow fever amongst human communities both on the American and African continents. The tree-living species suggested by Lumsden and Buxton were *A. africanus*, *A. luteocephalus* and *A. Simpsoni*. In an outbreak of yellow fever in the Sudan, Lewis *et al.* (1942) suggested transmission by *A. taylori* and *A. natalensis*. Those species actually incriminated by virus isolation are *A. aegypti*, *A. simpsoni* amd *A. africanus* (Haddow, personal communication).

Yellow fever in the South American continent is undoubtedly a 'younger' disease, and monkeys, especially howlers, are often severely affected and large numbers die. A wide variety of mosquito species are concerned with its spread. These are chiefly *Haemagogus* spp., *Aëdes* spp. and *Sabethes* spp. It would serve no useful purpose to review in detail the number of species involved, but it is important to note that the spread of yellow fever into new areas and its survival in already infested areas depends to a great extent on the bionomics of the mosquitoes concerned with its transmission. Each species differs from others not only in its favoured habitat and its feeding habits, but also by the means which it has adopted for survival over the dry seasons which may be prolonged in some areas. In relation to yellow fever, the mosquito is normally regarded as the vector and the primate the reservoir. This view is erroneous as has been shown by Bugher *et al.* and subsequently discussed by Ross and Gillett (1950) amongst others. It is usually accepted that the blood of simians infected with yellow fever carries transmissible virus for no more than six days, after which immunity is developed. Nevertheless, Smithburn and Haddow (1949) found that occasionally monkeys circulate virus in considerable quantity for more than six days; examples are tabulated in their paper. *The reservoir of the virus is the mosquito.* The survival of the disease, therefore, depends on the survival of mosquitoes through unfavourable seasons and their ability to infect non-immune hosts and thence further mosquitoes during the life span of the insect. Mosquitoes are unharmed by the virus, which multiplies in them at a greater rate under hot conditions.

Anthomyidae

Glossina spp. (Tsetse-flies)

The chief role of the tsetse-flies in medicine is that of transmitters of trypanosomiasis, including sleeping sickness of man. In the life history of this disease, the role of the vertebrate reservoir, whether wild antelopes or reptiles, is recognised, but there is virtually nothing known about monkeys in this respect. The ground-living baboons are immune to human trypanosomes, and their blood serum carries a protective factor. All other species of African monkeys are believed to be susceptible. Little information likewise is available as to the extent to which any species of *Glossina* feeds on ground-living or arboreal primates other than man. Weitz (1963) gives the following percentage figures for blood meals identified from the stomach contents of some *Glossina*:

Glossina sp.	Man (%)	Baboon (%)	Monkey (*Cercopithecus* spp. *Erythrocebus* sp. *Colobus* sp. (%)
G. swynnertoni	3·4	0·2	0·1
G. austeni	5·9	—	—
G. fuscipleuris	0·5	0·2	—
G. tabaniformis	—	—	—
G. morsitans morsitans	11·1	0·5	0·1
G. morsitans orientalis	5·8	1·0	0·3
G. morsitans submorsitans	18·3	1·0	1·0
G. pallidipes	2·8	—	—
G. longipalpis	1·6	—	0·2
G. fusca	—	—	—
G. longipennis	0·3	—	—
G. brevipalpis	0·7	—	—
G. palpalis fuscipes	21·1	0·2	0·2
G. palpalis palpalis	35·8	0·7	3·4
G. tachinoides	54·1	2·1	—

The figures given by Weitz indicate that a number of species do prey on both monkeys and man, and the extent to which this occurs may vary in different localities. The figures for *G. palpalis palpalis* are disturbing, although monkeys have not so far been incriminated as vectors of *Trypanosoma gambiense* in any of the great epidemics of sleeping sickness. In addition, either monkeys or baboons could be reservoir hosts of *T. rhodesiense*, the causal organism of Rhodesian sleeping sickness, and Nash (1948) considers that in some areas, notably the Gambia, the baboon is an important host of *G. morsitans*. He writes in *Tsetse Flies in British West Africa* as follows:

To turn to baboons, in my fact-finding trip in 1945, I strongly suspected that baboons must be the main source of food for the enormous *G. morsitans* population in the Gambia.

There is some evidence for this suggestion. In the Mandingo language this species of tsetse is called 'the tsetse of the black monkey'. Simpson (1911) wrote:

Close to the town of Buiba, there is a large open clearing, and on approaching it we saw an enormous number of dog-faced baboons (*Papio sphinx*) of all sizes. The road ran through the middle of this area, and on my arrival the baboons scattered to both sides. While passing over the ground recently occupied by them, the pony was badly attacked by tsetse which had evidently followed the baboons to the open country; many of these were gorged with blood and could fly only with difficulty. On three occasions during my tour in the Gambia I have witnessed this phenomenon; namely the predilection of *Glossina morsitans* for following a troop of baboons, even in the heat of the day, from their shady retreats into the open.

Nash states, however, that there is no evidence that tsetses feed on monkeys in the upper arcades of the forest, and he quotes from a letter sent by him to Potts and published (Potts, 1955) about the rare *G. nashi*, which he supposes may be a denizen of the high canopy feeding on monkeys:

Mr Gabesin has not been relying on the tsetse attacking him, but has been employing my technique of searching for the inactive flies at rest on saplings, creeper stems, etc. Hence the apparent rarity of the species is not solely due to its dislike of man as a host. Possibly this species rests so high up that it is normally inaccessible to a catcher, and only comes down to feed on hosts other than man; or it may be a denizen of the high canopy, normally feeding on monkeys and rarely descending to ground-level.

Nash adds that only six specimens of this fly have been captured in Nigeria, but he cannot believe it to be so rare as would appear. He believes that this tsetse and some of the other rare members of the *fusca* group may have a tree-top habitat in the high forest. Certainly many members of this group have been seen resting at heights which defied capture, say at twelve feet and above.

There is abundant evidence that tsetse flies could carry disease from arboreal or ground-living primates to man, and vice versa, but little evidence that they do so. There is some suggestion that certain tsetse-flies may be upper-canopy feeders, which occasionally descend to ground level, but there is little precise knowledge about this; ecologically it seems unlikely that tsetse-flies have not derived species to fill this niche.

Diptera (general)

Little is known of the extent to which arboreal primates suffer from other dipterous predators, nor whether they transmit disease and, if so, whether such diseases are transmitted to man. Most Diptera rely on water (e.g. Tabanidae) or decaying vegetation or dung (e.g. *Stomoxys*) for breeding, and such conditions would not be expected in the upper arcades of the forest – or so one supposes, in ignorance of possible conditions for breeding in holes in the trees, or amongst decaying epiphytes. Monkeys are frequently hosts to various genera of filaria, some of which are transmitted by mosquitoes, some by other Diptera. These are especially frequent amongst American species, where there must be a high vector-infection rate. The same is true of South American species of birds and reptiles. It has been suggested that baboons and mandrills may be reservoirs of the human filaria, *Loa loa* (Duke, 1957) which is transmitted by dipterous flies belonging to the genus *Chrysops*. A similar filaria is found in chimpanzees and gorillas.

The bionomics and infection rate in the Cameroons of species of *Chrysops* – *C. silacea*, *C. dimidiata*, *C. langi* and *C. centurionis* – have been described by Duke (1954, 1955a, b, c, d, 1958, 1959 and 1960). Duke's studies leave little doubt that *Chrysops* spp. are transmitters of *L. loa* both to monkeys and man; they live and bite at all canopy levels in the forest, and are attracted (*C. silacea* and *C. dimidiata*) from the upper forest layers to ground level by the smoke of wood fires, thus detecting and attacking humans with business in the forest. In relation to the transmission of disease between monkeys and man under feral conditions, *Chrysops* is plainly an important agent.

Flies of the genus *Hippobosca* are reported as predators of monkeys and lemurs, as well as man, but are not associated with the spread of disease.

Trapido (private communication) has found that in Panama a species of *Phlebotomus*, common in canopy conditions, attacked humans used as bait during the evening. These would presumably feed normally on sleeping arboreal monkeys. The species concerned is involved in the transmission of human leishmaniasis.

That midges (*Culicoides* spp.) are habitual predators of monkeys in Africa is shown by the frequency with which monkeys are infested with a parasite which only they transmit, *Hepatocystis kochi* (discussed later). *H. kochi* is not a parasite of man.

5 Endoparasites

Trematoda (Flukes)

The arboreal primates, owing to their habitat, are improbable hosts for most groups of trematodes. Nevertheless, reports of infection by some species are available. In the case of the schistosomes, the monkey has been used by a great many workers as an experimental animal, and experiments have shown it to be susceptible to infection by all species which cause disease in man.

Amphistome Flukes

Watsonius watsoni Deschiens (1940) discovered the amphistome fluke *Watsonius watsoni* at the Institut Pasteur in Paris in four baboons (*Papio cynocephalus*), newly imported from the former French Guinea. Pick (1951) described a new species from the same host, which he named *W. deschieni*. Flukes were present in the intestines of the affected baboons in very large numbers and caused diarrhoea and cachexia, followed by dehydration, wasting and death. These authors considered that infection might be carried for many months before clinical signs appeared; and often the disease might not progress beyond the stage of light diarrhoea. Stiles and Hassall recorded *W. macaci* for *Macaca irus*, but not for man, *W. watsoni* for *Cercopithecus callitrichus*, and for man. The life history of this parasite is unknown; and it is also not known whether it has any importance in the health problems of human populations in places where monkeys are infected.

Dicrocoelid Flukes

1 *Athesmia foxi* Stunkard (1923) recorded the recovery of twenty-six flukes from the hepatic and interlobular bile ducts of *Cebus apella* and (1928) described the fluke. This parasite is also

recorded by Stiles and Hassall from *Callicebus* sp. in Brazil. There is no record of infection in man.

2 *Eurytrema* spp. *Eurytrema brumpti* was reported from the Congo in *Pan troglodytes* by Railliet, Henry and Joyeux (1912), and *E. setoi* was found in *Macaca irus* in Japan by Kobayashi (1921). It was also reported by Stunkard (1949) from a gorilla, and by Stunkard and Goss (1950) in a young coast gorilla, newly imported into New York from Africa. There are no reports of cases in man.

3 *Controrchis biliophilus* This fluke was recovered from the gall bladder of *Ateles geoffroyi* at the Washington Zoo by Price (1928). Ruch supposes flukes of this type to be common parasites in the gall bladders of monkeys, and quotes Bablet *et al.* (1951) who found members of the same family in *Papio cynocephalus*. There are hitherto no records in man.

4 *Dicrocoelium* spp *Macaca speciosus* [?*silenus*] was found by Kobayashi to harbour *Dicrocoelium macaci*, but there are no records of this species in man, although he has been found to be the host of *D. dendriticum*.

Schistosomatidae

Stiles and Hassall cite references to *Schistosoma magna* as a parasite of apes and *Cercocebus torquatus*. This parasite does not appear to have been reported from man and is therefore not the cause of a zoonosis. It appears, however, that most simian primates can be experimentally infected with the human species of *Schistosoma*, namely *S. mansoni*, *S. haematobium* and *S. japonicum*. Cameron (1928) found five of eight *Cercopithecus sabaeus* monkeys infected with *S. mansoni* on the island of St Kitts in the West Indies. On this island the monkeys were introduced and presumably had become infected as the result of contact with slaves from Africa. Nelson, however, quotes a private communication from Shaw (1959) who stated that schistosomiasis was no longer a problem in St Kitts, and nothing was now known about the monkey infection. This is strange, because Jones (1932) considered that *S. mansoni* was a major public health problem on the island.

Little attention was subsequently paid to the possibility that simian primates could be important reservoir hosts of schistosomes until Miller (1959–60) reported the presence of *S. mansoni* infection in 32 of 134 baboons captured near Kibwezi in East Africa. These findings have been further reported by Strong, McGill and Miller (1961), Strong, Miller and McGill (1963). These workers, who discovered the high rate of infection in baboons during a survey of atheroma in wild baboon communities, considered that the baboon must be regarded as a reservoir host of this schistosome.

As a result of this work a survey of schistosome infections as zoonoses in Africa was made by Nelson and reported by him (1960, 1965). His very able (1960) review should be studied in the original, since he has been at pains to separate conflicting reports, some of which appear to arise from faulty identification of parasites or their eggs. Nelson recognises that other simian primates can be infected with schistosomes, but points out that baboons are more likely to be exposed to infection because they are largely terrestrial, like to reside near snail-infested water, and drink at water-holes, where they are frequently seen, apparently searching for snails on the underside of vegetation.

S. mansoni is normally transmitted by the snails of *Biomphalaria* spp. and baboons were only found infected with *S. mansoni* where *B. pfeifferi* existed and in addition where the resident human population was also infected. Baboons from regions without human populations were not found to carry infection. *Cercopithecus* monkeys from the same areas were found to be uninfected, except for a few specimens which carried only a very light load of schistosomes.

S. haematobium was also found in four of eight baboons examined, but these were only found in areas where the snail transmitter of this schistosome, *Physopsis* spp., was present.

Infections in the baboon resembled mild infections in man. The adult worms were to be found in the portal vein, including the intra-hepatic branches; eggs were present in both the liver and bowel, encysted in pseudo-tubercles. These were readily examined by crush techniques described by Nelson, less readily in material prepared for histological examination. Unaffected portions of the liver did not show abnormalities, and there was no cirrhosis. Nelson, in general, does not agree with the findings of Miller and his associates that the baboon is the reservoir host. As baboon infections are only present where water-holes are shared with human beings, he believes that the baboons acquire their infection from human sources and not vice versa.

General

Since all species of simian primates appear to be susceptible to a wide variety of trematodes, as indeed also is man, there always exists the possibility that infection may pass from one to another. That this could happen in the case of captive simians is improbable because of the life history of the parasites, though the possibility might exist with free-ranging colonies. The bionomics of simians makes it unlikely that they will, in most cases, be a serious source of infection in the wild. It is noteworthy that the baboon appears to be the most commonly affected animal apart from man, no doubt because of its ground-living habits. The commonest trematodes to affect tree-living monkeys are the Dicrocoeliidae, the second intermediate host being an ant which may well ascend into trees and be eaten by the monkeys. The zoonoses of this group of flat worms appear, therefore, to be associated to a great extent with the habits of host, parasite and vector.

Cestoda (Tapeworms)

Simian primates frequently carry the larval stages of Cestoda of a wide variety of species. The adult tapeworms carried are mainly *Bertiella* spp. (Anoplocephalidae), acquired by the ingestion of free-living mites. Simians do not naturally become infected with human tapeworms of the genus *Taenia*, which are acquired by consumption of the cysticercoid stages in the flesh of pigs or oxen.

Diphyllobothriidae

Sparganum Sparganum is the name given to the larval form of the tapeworm *Diphyllobothrium*. Infection in man and Felidae is acquired by consumption of the larvae in fish, these having acquired infection from the consumption of the first-stage larvae in fresh-water crustacea (*Cyclops*). Ruch describes cases of experimental infection of monkeys with spargana, which are pearly-white, worm-like organisms a few millimetres to several centimetres long.

Occasionally, spargana are found in wild monkeys also, although they do not often cause symptoms of disease. Sometimes, however, the parasites cause a disease in monkeys which is similar to elephantiasis in man. Stiles and Hassall quote a number of references to this infection by a variety of *Sparganum* spp. in man, baboon (*Papio hamadryas*), *Macaca nemestrina*, ape, *Saimiri sciureus* and *M. speciosa*. Spargana may be found in the skin and muscles, especially the deep abdominal muscles, the tongue, the glottis, the diaphragm, the upper arms and the thighs.

The position of simians as intermediate hosts of *Diphyllobothrium* spp. does not receive attention in the literature. It is thus not known whether sparganosis in monkeys is accidental or whether it is a stage in the life cycle of the parasite leading to carnivorous Felidae and to man, in those areas where monkeys are eaten.

Diphyllobothrium Ruch states that for obvious reasons no cases of infection of monkeys by *Diphyllobothrium* have been reported. Nevertheless, Stiles and Hassall have recorded two cases: one of a monkey (q.v.), the other of *Macaca fuscata*. The fishing habits of certain macaques (e.g. *M. cyclopis* in Formosa) should not be overlooked, when disease cycles are considered. Furthermore, the existence of sparganosis in mammals provides a possible channel of infection for any of the primates, such as chimpanzees, which occasionally add a meat dish to their diet.

Cyclophyllidea

Raillietina spp. though normally parasites in lower animals, are occasionally found in man and a few cases are reported in simian primates. Ruch, for instance, states that at least two species, *R. alouattae* and *R. multitesticulata* occur in South American monkeys. He quotes Joyeux and Baer (1949) who believe that *R. alouattae* is identical with *R. demerariensis*, which affects humans living in the same area; they suggest that the monkeys are the natural hosts and reservoir of infection.

Hymenolepidae

Stiles and Hassall catalogue a number of *Hymenolepis* spp. as parasites of man and monkeys – *Callicebus personatus* in Brazil, *Cercopithecus nictitans* in West Africa and *C. smithi*. The two most important species, *H. diminuta* and *H. nana*, affect both man and Old-World monkeys; *H. nana* may also infect mice, which could well convey infection from forest-living primates to human settlements. Infection may be spread either by faecal contamination of food or by fleas and beetles. *H. diminuta* is a parasite in rats and mice, and is spread by butterflies and moths as intermediate hosts. Human infections have been traced to cereals infested with meal moths and meal worms. Infection of insectivorous species of monkeys can, of course, easily occur.

Anoplocephalidae

The chief tapeworm in apes and monkeys, parasitic in the adult stages, is *Bertiella* spp., and this will sooner or later be encountered by all concerned in the care of simian primates. These tapeworms can also infect human beings, being chiefly found in children. They are also found in dogs. Infection probably occurs by the ingestion of free-living mites, though the life history is not known with certainty. Children usually become infected in areas where monkeys are common and are kept as pets. Symptoms are not usually serious in primates, including man. Stiles and Hassall give the following catalogue of recorded infections:

Bertiella cercopitheci	*Cercocebus torquatus*
B. conferta	*C. callitrichus*
	Macaca irus
B. fallax	*Cebus capucinus*
B. mucronata	*Homo sapiens*
	Pan sp.
	Cercopithecus sp.
	Callicebus personatus

B. polyorchis	*Macaca irus*
B. satyri	*Homo sapiens*
	Pongo pygmaeus
	Hylobates hoolock
Bertiella studeri	*Homo sapiens**
	Pan troglodytes
	Pongo pygmaeus
	Hylobates hoolock
	Papio hamadryas
	Macaca irus
	Cercopithecus pygerythrus

* See Roy (1938), *Homo sapiens* is not given as a host by Stiles and Hassall.

This catalogue is to be taken as a sample of hosts and species only, since the tapeworm is evidently well adapted to simian primates, and one must expect infection to be found in any species in any part of the world. It is more frequently found in the Old World than in the New, but is present in both. Its chief danger to man arises when captive monkeys play with children.

Moniezia A species of Anoplocephalid parasite occasionally found in South American monkeys is *Moniezia rugosa*; primarily a parasite of ruminants. This does not appear to affect man, but is reported from *Brachyteles* sp., *Ateles* sp. and *Alouatta caraya.*

Taeniidae

A characteristic of the Taeniidae is that the adult worms are in most cases less serious parasites than the larvae. Simian primates are rarely infected by the adults, but cases have been recorded of infection with all groups of larvae which occupy diverse sites—some causing serious, if not dangerous, disease. The following table classifies adults and larvae:

Adult	Larval form
Taenia	*cysticerus*
Multiceps	*coenurus*
Echinococcus	*hydatid*

The first two genera are morphologically similar, except in the larval form, in which *Multiceps* produces many heads (*coenurus*) whereas *Taenia* remains single (*cysticercus*). The danger from *hydatid* cysts arises as new larvae bud from a germinal layer of the cyst. Because of this the *hydatid* may grow to a large size in the parasitised organ and endanger life.

Graham (1960) draws attention to the occasional presence in monkeys of larval cestodes of a distinctive type, allocated to the genus *Tetrathyridium*. These are assumed to be the larval stages of tapeworms of the genus *Mesocestoides*, the adults of which are found in the intestines of various carnivores. These cysts do not appear to be dangerous and are often destroyed by phagocytes.

Ruch quotes a case reported by Walker (1936) in which the *cysticercus* of *Taenia solium* was found widely distributed in the organs of a monkey, including brain, heart, liver, spleen and both kidneys; they caused little harm and virtually no reaction. He also quotes the case of cysticercosis in a baboon described by Spitzker (1878), in which parasites were present in brain and liver.

Coenurosis has been described in cercopitheques and baboons. The brain is sometimes affected, but more common sites appear to be the musculature and abdominal cavity. Cysts are often present in gelatinous masses. The adult appears to be the dog tapeworm, *Multiceps serialis*, though the dog can hardly be the host in wild conditions.

Dumas (1953) – as quoted by Ruch – described *hydatids* caused by larval *Echinococcus granulosus* in lemurs, monkeys and baboons. A number of other cases of cysticercosis, coenurosis and hydatid disease have been described by various workers, though some appear to use the term 'hydatid' loosely to indicate tapeworm cysts generally, thus making it difficult to interpret the situation correctly.

Stiles and Hassall give the following host records:

Cysticercus cellulosae	*Homo sapiens*
	Ape and monkey spp.
	Macaca sylvana
	M. silenus
	Cercopithecus cephus
	Erythrocebus patas

Cysticercus tenuicollis	*Homo sapiens* *Mandrillus sphinx* *Pygathrix* sp. *Semnopithecus* sp. *Macaca sylvana* *M. silenus* *Cercocebus* spp. *Cercopithecus* spp. *Erythrocebus patas*
Coenurus (*M. serialis* larva)	*Homo sapiens* *Theropithecus gelada* *Macaca* spp. *Lemur catta*
Echinococcus granulosus (hydatid)	*Homo sapiens* *Papio ursinus ursinus* *Theropithecus* sp. *Macaca sylvana* *Lemur macaco.*

Apes and monkeys, therefore, can harbour the larvae of human tapeworms, but this is of little significance as a zoonosis except amongst rare and primitive monkey-eating communities. The most important of these larvae is the *hydatid*, for which, under wild conditions, simians can serve as a reservoir; where monkeys are kept in captivity, they are unlikely to be eaten by dogs so that spread to a human host by this route is improbable.

Acanthocephala (Thorny-headed Worms)

Various genera and species of Acanthocephala are parasites in most groups of monkeys and apes, and may cause serious symptoms. Many are transmitted by cockroaches, and possibly other insects, by ingestion. No cases have so far been described in man, probably because of his distaste for cockroaches.

Nematoda (Roundworms)

Strongyloidea (Hookworms and Nodular Worms)

The main species of these worms affecting simian primates are: the nodular worms *Oesophagostomum* spp., the human hookworms *Ancylostoma duodenale* and *Necator americanus*, and *Ternidens deminutus*.

Oesophagostomum spp. By far the most important of these worms in simian pathology are the nodular worms, known as oesophagostomes, which are common parasites in herbivorous animals. They do not commonly occur in man, but infestation can be contracted from infected apes and monkeys. Unlike the hookworms, infection with *Oesophagostomum* spp. occurs by ingestion, the larvae hatching on the ground. The eggs are indistinguishable from hookworm eggs, and thus human infection may often be mistaken for the latter.

In either man or monkeys, light infections probably do little harm. Infections, however, are often heavy and may be massive, in which case the worms may cause a serious condition which may lead to death; at least one known human case has proved fatal (Chandler, 1955). The worms commonly infect the large intestine and burrow beneath the mucous membrane to form the characteristic nodules. These are black, rounded bodies, which contain motile worms and dark-coloured blood. Dr L. G. Goodwin (personal communication) states that, in man, large nodules have occasionally been mistaken for neoplasms, and the correct diagnosis only made by colonic resection. Oesophagostomiasis can be a very serious disease responsible for the deaths of newly captured anthropoid apes, particularly gorillas, but affecting also orang-utans and chimpanzees; strangely there are very few reports of it in gibbons. Otherwise, all groups of Old-World monkeys, including baboons, may be found affected. It is not found in New-World monkeys.

The oesophagostomes are thus dangerous parasites in the lower primates, and they may be transmitted to man either from wild monkeys or from animals in captivity. The main species are *O. apiostomum* and *O. stephanostomum*. The latter species is found in wild gorillas and has been blamed by Georges Sabater Pi (1963) for the deaths of as many as 80% of young wild gorillas after capture in Rio Muni, Spanish Guinea.

Ancylostoma sp., *Necator* sp. and *Ternidens deminutus* Stiles and Hassall quote infection of primates of various species with both *Ancylostoma* and *Necator*. *A. duodenale*, the common human species, has been recorded from man, chimpanzee, gorilla, gibbon and various macaques. Other species are quoted from the slow loris and the South American howler monkeys, but cases of infection in man with these species are lacking.

Necator americanus is described from man, chimpanzee, gorilla and the patas monkey. Certain other species have been taken from gorilla and chimpanzee, but these have not been reported from man.

Ternidens deminutus has been reported from man, cercopitheques and macaques by Stiles and Hassall; Amberson and Schwarz (1952) have reviewed the incidence of this parasite in man and other primates. This parasite is intermediate between the hookworm and the nodular worm in the sense that it causes nodules by burrowing into the wall of the caecum, but the nodules do not contain worms. The worms resemble hookworms in that they ingest blood and cause anaemia. They commonly infect Asian and African monkeys and also human beings resident in Africa.

Eunematoda

Nematodes of this family have both free-living and parasitic phases. They are mostly harmless parasites in sheep and other ruminants, but infection in monkeys, apes and man can at times occur and be serious.

Strongyloides The classification of the genus *Strongyloides* taken from primates, including man, is confused. There are, however, at least three species, *S. cebus* and *S. fuelleborni* from New- and Old-World genera of primates, and *S. stercoralis*, which is a parasite in man and may cause serious symptoms. The latter species, however, is certainly also a parasite in gibbons and chimpanzees, and infections with it are, therefore, zoonoses. *Strongyloides* is a very common parasite in all groups of monkeys and apes.

The worm is small (2–4 mm long by 40–50 μ in diameter); it has alternate free-living and parasitic generations. The fertilised eggs produce *rhabditoid* larvae which feed and grow in the free state until they metamorphose into long *filariform* larvae; these penetrate the hosts' skin and thus start the parasitic stage. The *filariform* larvae enter the blood stream and pass to the heart and lungs, where they break out into the alveoli and develop further. Most then ascend the respiratory tract, are swallowed, and reach the intestines.

The female becomes fertilised during the course of these migrations, and upon reaching the intestines she burrows into the mucosa and begins to deposit eggs. When the eggs are laid and the larvae develop the cycle is repeated. Sometimes, the *rhabditoid* larvae metamorphose into the *filariform* larvae in the bowel, and they can then renew the cycle in the same host.

These worms cause extensive tissue destruction and haemorrhage both in the lungs and in the bowels. Sometimes, infection causes a mild diarrhoea but it may be so severe as to cause haemorrhage and death. Eosinophilia is commonly associated with infections by these worms. Symptoms and lesions in both man and the lower primates are very similar.

A fatal case of strongyloidiasis in an immigrant negress in Britain was described in detail by Wilson and Thompson (1964). Strongyloidiasis is endemic in tropical and sub-tropical countries, and the results may at times be serious. The case described in the paper referred to was due to *S. stercoralis*, larvae of which had penetrated the mucosa and sub-mucosa of the wall of the lower bowel causing widespread and serious lesions. As a consequence, the patient had developed pyaemia due to *Escherichia coli* and succumbed to purulent meningitis.

Other Nematode Infections

The species of whipworm (*Trichuris trichuris*) found in apes and monkeys is similar to that found in man and probably identical. Fatalities have been described in monkeys and chimpanzees, but this infection is not usually serious.

Capillaria hepatica is a slender worm which inhabits the substance of the liver and has been described in a number of species of lower primates. Infection in monkeys can cause fatal hepatitis. The worm is also found in wild rodents. Transmission occurs by the ingestion of an infected liver. Since human cases can occur (MacArthur, 1924), cross-infection probably takes place where monkeys or rats are consumed as articles of diet.

Enterobius vermicularis, the irritating pinworm found in man, has been found in captive apes, gibbons and marmosets, and can be passed from man to monkey and vice versa.

Ascaroidea

Ascaris lumbricoides only very rarely infects monkeys and apes, and cross-infection is not a serious danger. In general, ascarids are rare in monkeys, although they may become infected by *Toxocara canis* and *T. cati* also occasionally found in human patients. *Toxocara* is important, first because of its unusual life history, secondly because the second-stage larvae cause the disease 'visceral larva migrans', which is a serious hazard to children.

Toxocara canis and *T. cati* are almost universal parasites in adult dogs and cats respectively, but normally the infection is 'silent', the adult worms living inactive lives in the tissues. In the case of bitches, adult worms appear in the bowel within twenty-three to forty days of whelping, and eggs are passed in the faeces at this time. Puppies suffer from pre-natal infection and pass eggs in the faeces for some two months after birth. Thus both puppies and bitches, prior to whelping and when nursing their puppies, constitute a hazard to children if care is not taken to see that fingers and food do not become contaminated. Infection will result in a condition of 'visceral larva migrans', which leads to persistent eosinophilia, intermittent fever, loss of appetite, failure to gain weight, cough, muscle and joint pains, abdominal pain or neurological symptoms. Sometimes, larvae penetrate the eyes, and loss of sight occurs, as a result of damage to the retina. The condition is sometimes fatal, but usually passes by natural recovery with time.

The life cycle was first described by Sprent (1956, 1958) and has been reviewed by Gibson (1960). There is, however, no information on the bionomics of adult *Toxocara* in humans or simians, so that the danger of a monkey pet being infected and transmitting 'larva migrans' to children cannot be assessed. One would suppose that the main danger, both to children and to monkeys, lay in an infected dog or cat.

Spiruroidea

A number of genera of the Spiruroidea have been described from a number of species of monkeys. These worms are transmitted by the ingestion of eggs by arthropods, which infect a new host when they themselves are eaten. Leiper (1911) states that *Physaloptera caucasia*, a monkey parasite, is occasionally seen in African natives.

Filarioidea

Both adult filariae and microfilariae have been found *post mortem* in both apes and monkeys, though no case of elephantiasis associated with this cause has been described. The adult worms dwell in the lymphatic system or in the liver, and may be recovered from many different sites, especially the peritoneal cavity and often the tendon sheaths of the wrist. They are often present in large numbers, though only one or two worms may be found. Incidence is particularly high in South American primates, and also in South American birds and snakes. Microfilariae may be found in blood smears and skin scrapings, but there is in some groups (excluding *Dipetalonema*) a diurnal variation in their presence.

The commonest filariae in primates belong to the genus *Acanthocheilonema* (*Dipetalonema*). The rate of infection is high, particularly in wild and newly caught specimens. Apart from virtually all species of South American monkeys, infection has been described in gorillas, chimpanzees, baboons, cercopitheques and macaques. Infection in man is recorded by Stiles and Hassall, and Manson-Bahr (1960) records *Dipetalonema perstans* (under the generic name *Tetrapetalonema*) in man. The adult worms occur in the peritoneal or pleural cavities (but rarely in the pericardium) of man and apes in tropical Africa generally and in North Africa, as well as in the entire South American continent, but they are not very pathogenic. This species shows no periodicity in the blood. *D. streptocerca*, however, the adults and microfilariae of which have been found in the skin of man in Central Africa and Ghana, may cause oedema and elephantiasis of the skin.

Dirofilaria sp. has been described from the right heart cavity in orang-utans and *D. corynodes* occurs in the monkeys (*Cercopithecus aethiops* and *Colobus* sp.); neither of these species have been recorded from man.

The genus *Loa*, which is highly dangerous to man, causing eye and skin lesions, has been found in a number of Old-World monkeys. It appears to be well adapted to baboons and causes little harm to them. This filaria is transmitted by the biting fly *Chrysops*, and there are therefore greater opportunities for cross-infection than with the mosquito-transmitted filariae. Possibly, the baboon is a reservoir of infection, and Sandground (1936) suggests that the eye localisation and subcutaneous wandering of *L. loa* in humans are evidence that man is an alien host. The same author believes that *L. papionis* of the baboon may not be a valid species, but identical with *L. loa*. In monkeys, the adult worms were found in subcutaneous tissues, pericardium, diaphragm and in the peritoneal cavity. This filariasis is not very pathogenic, and worms were not found in aberrant sites.

Amongst macaques in South-east Asia, many microfilariae have been found and described, some of these are probably the larvae of *Wuchereria malayi*. Together with *W. pahangi*, and *W. patei*, these larvae belong to the recently established genus *Brugia* (Buckley 1960), and all of them are parasitic in the lymphatic systems of primates. *W. malayi* is a dangerous parasite in man, but *W. pahangi* and *W. patei* have only been reported from simians and pro-simians.

W. bancrofti, whose microfilariae can be distinguished from those of *Brugia*, infects the lymphatic system of man in the tropical regions of both the Old and the New World, but has so far not been recorded from simian primates.

Onchocerca spp., transmitted by the black fly *Simulium damnosum*, cause fibrous nodules in the subcutis in man and are associated with 'river blindness' in West Africa. They have not been found in lower primates, although Duke (1962) in the Cameroons succeeded in infecting chimpanzees artificially.

A young female chimpanzee, kept in a screened cage at the Helminthiasis Research Unit at Kumba from March 1960 onward, was inoculated subcutaneously and intradermally by syringe with a total of about 280 infected larvae of *Onchocerca volvulus* extracted from 'wild' *S. damnosum* which had been fed on heavily infected human volunteers. Skin snips were taken at regular intervals from the chimpanzee's head, shoulders and legs – the regions inoculated – and in February 1962, after about twenty months, microfilariae which could be identified morphologically as those of *O. volvulus* were recovered. No adult worms, no nodules, skin or eye lesions had been observed at the time of publication, during the first half of 1962, and the animal was kept alive in order to observe the progress of the infection.

Between the time of inoculation and recovery of *O. volvulus* microfilariae occasional specimens of other microfilariae were detected, some resembling those of *Dipetalonema streptocerca* and very rare ones similar to those of *D. rodhaini*, both these, however, were considered to be natural infections, which have also been observed in six other young chimpanzees kept at Kumba.

This is believed to be the first recorded instance of successful transmission of *O. volvulus* to an experimental animal; attempts to transmit this microfilaria to other animals (seven drills, one mangabey, one Preuss' monkey, one goat) failed, though Caballero and Barrera (1958) mention natural infection with a species of *Onchocerca*, probably *O. volvulus*, once in a spider monkey.

This evidence shows that chimpanzees are possible carriers of *Onchocerca* sp. and could act as a reservoir for human infection.

To sum up, the filariae in company with other mosquito-transmitted diseases are ubiquitous and common parasites of all groups of apes, monkeys and pro-simians. To what extent they cause disease and loss in wild individuals is unknown, but particularly in New-World species they can cause serious losses of those brought into captivity. The microfilariae appear to be associated with signs of anaemia, and the adult worms frequently cause peritonitis or subcutaneous abscesses. Neither in captivity, nor in the wild, are the opportunities for cross-infection with man extensive, and one may suppose that infection is usually avoided both because of differences of habitat and because the chief infecting species have diverged. There exist, however, chances of cross-infection between baboons and man by *Loa* spp. in Africa, and in the Far East between macaques and man with *Wuchereria malayi*. The possibility exists that monkeys may be an important reservoir of *L. loa* transmitted by *Chrysops* spp. The position is reviewed on pp. 47–48.

Helminths—General

Generally speaking, man and the non-human primates have diverged widely with regard to the helminth parasites by which they are attacked. The main infecting species of worms are different in each case and the opportunities for cross-infection are not very great. Nevertheless, most worms retain the ability to infect both types of host and so constitute a potential danger, particularly where monkeys or apes are kept as pets. In this way the common oesophagostomes of monkeys can infect humans, and the hookworms of man can infect monkeys. The whipworms and pinworms, which are of little pathogenic significance, have diverged to a lesser extent and usually infect both groups.

The position of the helminths is in direct contrast to that of the pathogenic Protozoa of the bowel, which will be considered in the next chapter.

6 Protozoa

The parasitic Protozoa belong to four classes:

1 *The Sarcodina*, with their sub-class Rhizopoda, of which the amoebae are the most important members;

2 *The Mastigophora*, or *Flagellata*, such as the trichomonads, leishmanias and trypanosomes;

3 *The Ciliata*, of which *Balantidium coli* is the most important;

4 *The Sporozoa*, comprising the great group of non-locomotory parasites, including malaria.

It is most convenient to follow the method used by Ruch in discussing first the intestinal Protozoa, which comprise a heterogeneous group containing the amoebae, some flagellates, some ciliates and the Coccidia from amongst the Sporozoa.

The Intestinal Protozoa

The intestinal Protozoa have diverged far less amongst the primates, including man, than have most other groups of parasites. Wild primates commonly carry amongst their intestinal fauna a number of potentially harmful and harmless Protozoa, some of which may become active and cause disease symptoms under conditions of stress associated with capture.

The amoebae are of great importance, but their story is simple, and little need be said about them. *Entamoeba coli*, normally a harmless commensal, is commonly present in Old-World primates, including man, but has only been recorded once in New-World species by Dobell (1936) from a marmoset, *Hapale jacchus*; he experimentally infected man with *E. coli* from this marmoset and also from a macaque. It appears, therefore, that the simian strain can infect man and vice versa. *E. histolytica* is harmless to Old-World monkeys (macaques) but under experimental conditions, produced clinical signs in New-World species (Hoare 1963). This is a dangerous parasite in both man and the anthropoid apes, and the term 'amoebiasis', or amoebic dysentery, is generally restricted to mean infection with this parasite. It may, however, be present without causing serious effects. In both man and monkeys there appear to be different strains of *E. histolytica*, though it seems certain that most can be transferred from the one to the other.

Affected apes or monkeys, when kept in captivity, may be a danger to man; it is hardly to be supposed that cross-infection can occur to any great extent under wild conditions. In either host, ulceration of the colon may occur with diarrhoea and dysentery; both hosts are liable to be affected with abscesses of the liver which may be fatal. A number of other species of *Entamoeba* and *Endamoeba* are listed as parasitic in man and simian primates, both by Ruch and by Stiles and Hassall; some are said to be pathogenic to a greater or lesser degree, but little is known about them.

The intestinal flagellates found in the lower primates are mainly the same as those found in man and must be supposedly readily transmissible from the one to the other. Those listed by Ruch are: *Giardia intestinalis*, *Trichomonas* spp., *Chilomastix mesnili*, *Embadomonas intestinalis*, and perhaps *Enteromonas*. *G. intestinalis* is a parasite of low pathogenicity in both groups and is easily cured. It is found mostly in children, and there is evidence that it is harmful to chimpanzees only if they are very young. The main symptom is diarrhoea with some manifestations of colic.

Trichomonas is frequently found during faecal examinations of chimpanzees and other simian primates during the course of examination for amoebae or other parasites. It appears to be quite harmless to these hosts, although no doubt cross-infection between man and monkey could occur. The common *T. foetus*, which causes infertility and abortion in domestic animals, can infect some of the lower primates, but not the human female. The venereal *T. vaginalis* of man has been isolated from both wild and captive monkeys, and macaques have been artificially infected with human strains of the parasite. Symptoms caused in both man and the lower primates are mild.

The most important of the ciliates in simian primates is *Balantidium coli*, a non-pathogenic bowel parasite of the pig. This parasite is commonly found in the great apes, in which it may be present for long periods without causing symptoms. In conditions of stress associated with captivity it may, however, cause serious ill-effects. Chimpanzees, for example, suffer from diarrhoea, colic and severe pains in the region of the caecum. Cysts are often not seen in faecal samples and a diagnosis of appendicitis is commonly made. There is usually accompanying anorexia, and death may ensue in a matter of weeks or months. At necropsy, dark patches may be seen in the colon and caecum, a centimetre in diameter, and histological examination of these reveals the cysts of *Balantidium*. *B. coli* is a rare parasite in man, but when present causes severe dysentery. There is little doubt that infection can pass from apes or monkeys to man and vice versa.

A very large and rather striking parasite of the great apes is *Troglodytella* sp. This is non-pathogenic and almost invariably present in the faeces of chimpanzees and gorillas. It has not been described from any other species and does not appear to infect man.

The Coccidia, of the family *Isospora*, are rare parasites in man, but they are non-pathogenic. Their role is identical in the lower primates. The genus, *Eimeria*, has not been described from primates including man, except for a single record of infection in a marmoset, *Hapale jacchus* (Hoare 1964).

The Mastigophora

Trypanosomatidae

Both *Trypanosoma gambiense* and *T. rhodesiense* can infect monkeys experimentally, and the course of the disease is identical with that in man. Symptoms caused by *T. gambiense* are relatively chronic and those by *T. rhodesiense* are acute. Baboons are said to be resistant, and baboon serum contains trypanocidal substances. There is no evidence of the extent to which the simian primates become infected with these trypanosomes in the wild, or whether they could act as reservoirs of infection. As has been seen, there is little information on the extent to which *Glossina* spp. prey on monkeys, other than baboons. Nevertheless, it is usually thought to be unlikely that simians could be a reservoir of *T. gambiense* sleeping sickness. The success in clearing areas of this infection by the removal of the human population suggests that there are no other susceptible animals, but the disease could reappear once human populations were reintroduced to cleared areas if a reservoir of infection existed in the monkeys. The position with regard to *T. rhodesiense* sleeping sickness is different, and monkeys could theoretically be reservoir animals, though ungulates are the usual reservoirs.

The argument given above in relation to *T. gambiense* sleeping sickness, although probably correct, is too facile to be accepted without question. A situation could conceivably exist whereby monkeys in the forest canopies regularly carried *T. gambiense* transmitted by tsetse flies living at high canopy level. In such circumstances, there could be cross-infection between canopy-dwelling animals and ground-dwelling animals at extremely infrequent intervals. Thus, the potential for sleeping sickness could exist in an area, unknown for decades except for sporadic cases in the human population. Such sporadic cases do unaccountably occur in forest areas. There is no evidence either to support or to refute such a supposition, but it is paralleled by the yellow fever story.

The common trypanosome of the pig, *T. simiae*, was so called because it was found to infect a monkey. There is no evidence either that it ever does infect monkeys in the wild, or that it can affect humans.

Infections with *T. cruzi*, known as Chagas' disease, occur in many Central and South American mammals, amongst them monkeys. It is possible that monkeys may be one reservoir of this disease, but there is no evidence that they are concerned in a disease cycle involving man. These trypanosomes are transmitted by triatomid bugs, including *Rhodnius prolixus*. The intermediate hosts pass faeces after a blood meal, and these may be rubbed into the eyes or sores around the lips of the main host; in this way the trypanosomes infect the human hosts. It is known that *Rhodnius* and possibly some other triatomes climb into palm thatch and into trees to hide and nest. In this way they may well parasitise monkeys; whether they in fact do so is unknown. Chagas' disease is serious in man, and there could be danger of cross-infection in places where suitable intermediate hosts exist.

T. rangeli is another trypanosome which commonly infects humans, dogs, cebus monkeys and opossums (*Didelphys*) in South and Central America (Venezuela, Colombia, Guatemala). This trypanosome is also transmitted by the triatomid bug, *R. prolixus*. It is easily distinguished from *T. cruzi* by its larger size. It is usually stated to be non-pathogenic, but some observers do not agree about this.

Leishmania Leishmaniasis is caused by three species of *Leishmania*:

1 *L. donovani*, prevalent in South America, Asia, Africa and the Mediterranean area, causes 'kala-azar'.
2 *L. tropica;* has a similar distribution to *L. donovani* and causes 'oriental sore'.
3 *L. brasiliensis*, confined to South America, causing the disease known as 'espundia', a widespread and distressing but curable disease in the Amazon Basin.

Simian primates can be experimentally infected with all three species of *Leishmania*, but they do not appear to cause natural diseases in wild monkeys, even though cases were quoted by Stiles and Hassall. The recognised reservoirs of these diseases are ground-living animals, chiefly gerbils, dogs, foxes and jackals, and transmission occurs by sandflies (*Phlebotomus* spp.), but see p. 48; Trapido believes that sandflies may attack at both treetop and ground level. Contacts with tree-living primates would appear to be infrequent, and monkeys have never been suggested by any workers as reservoirs.

Sporozoa

Plasmodium – Malaria

The malaria parasites of man and simian primates have diverged, so that each host is naturally affected by different species. In each, however, the parasites can be placed in similar groupings; in some cases the human strain from a particular grouping can infect monkeys, in other cases not – similarly with the infection of man by simian strains. The divergence of the simian and human parasites is indicative of the effect of the divergence of the two types of host into differing habitats over so great a period. Nevertheless, in some parts of the world – particularly Malaya – the ecological separation may no longer be so great as in the past, and there is a strong probability that malaria is an important zoonosis. Indeed some workers believe that malaria eradication schemes may be vitiated because human infection can be brought about by simian parasites.

The position with regard to malaria has been succinctly and excellently described by Hoare (1962). The following is reproduced verbatim with his permission:

Malaria Parasites

An interesting problem is presented by the malaria parasites of man and anthropoid apes. Thus chimpanzees harbour under natural conditions three species of plasmodia which, although morphologically indistinguishable from the human parasites, *Plasmodium vivax*, *P. malariae* and *P. falciparum*, have been given separate names, *P. schwetzi*, *P. rodhaini* and *P. reichenowi*, for the Benign Tertian (B.T.), Quartan (Q) and Malignant Tertian (M.T.) parasites respectively. The mutual relationship between the human and simian parasites was determined by cross-infection experiments, with the following results. It was demonstrated that the M.T. parasites of man and chimpanzee, *P. falciparum* and *P. reichenowi*, are strictly specific not only to their mammalian hosts but also to their anopheline vectors, and can therefore be regarded as distinct biological species. On the other hand, the B.T. and Q. parasites of man (*P. vivax* and *P. malariae*) and chimpanzee (*P. schwetzi* and *P. rodhaini*) are interchangeable between these two hosts (Rodhain, 1939; Rodhain and Dellaert, 1955; Bray, 1956; Garnham *et al.*, 1956). Hence it follows that the last two parasites are common to the hominid Primates. But, in view of the absence of close contact between these hosts, apes cannot be regarded as reservoirs of human malaria; it is more probable that man inherited these parasites from his anthropoid ancestors in the course of evolution.

There is increasing evidence that man is also susceptible to infection with the malaria parasites of lower monkeys. Thus, it has been known for some 30 years (Knowles and Das Gupta, 1932) that the quotidian parasite, *P. knowlesi*, could produce malaria in man, and at one time it was used in malario-therapy of General Paralysis of the Insane. It was also suggested (Anonym, 1961) that some B.T. infections of man in the Amazon region of Brazil might be derived from *P. simium*, a parasite of howler monkeys (*Alouatta*) which resembles *P. ovale* (Fonseca, 1951).

Quite recently more concrete evidence was produced that some forms of human malaria might be in fact zoonoses derived from monkeys. Thus, two years ago American workers (Eyles *et al.*, 1960) reported accidental laboratory infection of man in the United States with *P. cynomolgi* isolated from *Macaca irus* in Malaya, while later (Beye *et al.*, 1961; Coatney *et al.*, 1961) the monkey strain was transmitted through the bites of infected anophelines to a group of volunteers, who developed tertian malaria, which was passaged further through a series of volunteers. Though this parasite adapted itself to man, it remained infective to monkeys. Final proof of the infectivity of this parasite to man was provided by recent experiments (Contacos *et al.*, 1962) in which *P. cynomolgi* was transmitted with serial cyclical (*Anopheles freeborni*) and direct passages through 12 human volunteers producing in them typical tertian malaria.

It is now known (Wharton and Eyles, 1961) that the natural vector of *P. knowlesi* in Malaya is *Anopheles hackeri*, and that the local vector of human malaria, *A. babirostris*, also feeds on monkeys (Reid and Weitz, 1961), while the role of other Malayan anophelines (*A. maculatus*, *A. sundaicus* and *A. philippinensis*) was demonstrated experimentally (Warren *et al.*, 1962). Since certain malaria parasites are interchangeable between man and monkeys, it is conceivable that Malayans living in or on fringes of the jungle are frequently exposed to the bites of mosquitoes harbouring simian parasites, and become infected with monkey malaria.

These discoveries might have an adverse effect on the success of malaria eradication campaigns, for it will be necessary to consider the possibility that in enzootic areas man is liable to acquire monkey malaria. On the other hand, they may be of positive value in chemotherapeutic research, since for the first time malaria parasites of simians have become available for tests on antimalarial drugs [*i.e.* using natural hosts].

In a personal communication, Hoare (1964) states: 'In addition to the tertian parasite (*P. cynomolgi*) of Malayan macaques, it has recently been proved experimentally that the quartan parasite (*P. brasilianum*) of South American monkeys is cyclically transmissible to man. It is therefore possible that the disease in man infected with simian parasites is occasionally mistaken for human malaria.'

The classification of the simian malarias is confusing, but the following table will assist the reader to understand the position:

SIMIAN MALARIAS

		Man	Apes	Monkeys
Quartan Group		*P. malariae*	*P. rodhaini*	*P. knowlesi*
		(*P. knowlesi*)*	(*P. malariae*)	—
			P. hylobati	—
		(*P. brasilianum*)	—	*P. brasilianum*
Vivax Group (Benign Tertian)		*P. vivax*	*P. schwetzi*	*P. gonderi*
	I.	(*P. cynomolgi bastianellii*)	(*P. vivax*)	*P. cynomolgi*
				P. c. bastianellii
		(*P. schwetzi*)	—	*P. inui*
				P. osmaniae
				P. fieldi
	II.	*P. ovale*	*P. ovale*	*P. ovale*
		(*P. simium*)	—	*P. simium*
Malignant Subtertian		*P. falciparum*	*P. reichenowi*	

* Brackets indicate susceptibility of an unnatural host.

In the malignant sub-tertian group, no parasites are known except from man and the anthropoid apes. *P. falciparum* is unable to infect anthropoids, but it is not known whether *P. reichenowi* can infect man. The severe reaction caused in man by *P. falciparum* has suggested to many workers that it may be a rather new and ill-adapted parasite in man, a view supported by the absence of parasites of this group from the lower primates.

The vivax group is widely distributed throughout the world – Europe, Asia, Africa and South America and is represented by a large number of species. Cross-infection between man, apes and the lower simians has been established with some species, and both in Malaya and other far eastern countries representatives of the group may well be responsible for creating cross-infections between man and monkeys under natural conditions. The group must be regarded as consisting of ancient parasites of primates in the course of divergence.

Members of the quartan group occur in South America, Africa and Asia. Cross-infection can occur between man and apes in respect of *P. malariae*, and between man and monkeys in respect of *P. knowlesi* and *P. brasilianum*. Both in India and Malaya, it is believed that cross-infection with *P. knowlesi* occurs in natural conditions. It is unlikely that cross-infection between apes and man can occur in natural conditions in respect of *P. malariae* or *P. reichenowi*, because the latter is unable to infect *Anopheles gambiae*, and the former, by the same analogy, may be unable to infect the vector of *P. reichenowi*, whatever this may be. In South America, cross-infections involving *P. brasilianum* are a distinct possibility, but that they do occur under natural conditions has not been established.

Obviously, the bionomics and habits of the vectors of *Plasmodium* spp. are important, since a parasite must be capable of infecting both the human and simian host and also the vectors which attack each, or else both vectors must – even if infrequently – attack both man and simians. The extent to which these conditions are fulfilled in the case of the Plasmodiidae and Anophelinae is not known. It is only suspected to occur with any regularity in the cases of *P. knowlesi*, *P. c. bastianellii*, *P. simium* and/or *P. brasilianum*. These instances are sufficient to establish malaria as a potential zoonosis between monkeys and man in Malaya, other parts of Asia and South America. There is as yet no evidence that cross-infection between man and apes does take place in natural conditions or is of importance, but there is abundant evidence supporting the possibility.

The pathogenicity of malaria for simians is variable, but on the whole it appears to be well tolerated and not serious under natural conditions. *P. knowlesi* occurs naturally in *Macaca irus* and causes little harm; strangely it is very severe and often causes fatal disease in *M. mulatta. P. inui* and *P. cynomolgi* are very mild, and in the wild may exist in mixed infections of one or the other with *P. knowlesi. P. brasilianum* usually causes low-grade, non-fatal and long continued disease with relapses in its normal hosts, *Aotus* and *Hapale*; it may, however, be very acute in *Cebus* and *Ateles*. Laboratory simians are often found carrying latent infections which may flare in conditions of stress, trauma or ill-health of some other kind.

Other Sporozoa

A common parasite in African monkeys, closely allied to the plasmodia, is *Hepatocystis kochi*. The life cycle of this parasite is indistinguishable from *Plasmodium* except that it does not undergo schizogony in the erythrocytes to produce merozoites. The asexual cycle of reproduction occurs in the liver, and the parasite can therefore by definition not be classified with the plasmodia. The infection is symptomless in a wide range of monkeys, and no cases of human infection have been described. The normal range of malaria vectors and other mosquito species: *Anopheles maculipennis*, *A. quadrimaculatus*, *A. gambiae*, *Aëdes aegypti*, *Culex molestus* and *Taeniorhynchus africanus* cannot be infected with *H. kochi*, which is only transmitted by midges, *Culicoides* spp., as shown by Garnham *et al.* (1961). *H. kochi* is important because it led to the discovery of the extra-erythrocytic cycle of human malaria. It is extremely common and widespread in East Africa, occurring even on Nkose Island, an isolated, rocky islet in the vastness of Lake Victoria south of Sesse. This implies that monkeys must almost everywhere be bitten by *Culicoides* spp., which Garnham has shown to be the vectors.

Babesiasis is not uncommon in apes and monkeys, both in Africa and Asia. The disease caused may be serious, but there is no evidence that it can be transmitted to man. One of the infecting organisms that has been described in detail has been named *Babesia pitheci.*

Infections have also been described in lemurs, monkeys and apes caused by various organisms of indeterminate taxonomic position, e.g. *Anaplasma*, *Sergentella* and *Grahamella*. The only one of these which may be of significance in human disease is the latter, since this is closely related to *Bartonella bacilliformis*, the cause of Oroya fever. *Grahamella* is transmitted by sandflies, and cross-infection could conceivably occur in some places.

Toxoplasmosis appears to infect most warm-blooded animals, including monkeys and man. Natural outbreaks have, however, been mostly reported in New-World monkeys. Ratcliffe and Worth (1951) described five occurrences of toxoplasmosis occurring over a period of ten years in the small-animal house of Philadelphia Zoological Gardens. All affected animals, squirrel monkeys (*Saimiri sciurea*) and one spider monkey (*Ateles geoffroyi*) died; localisation appeared to be in the nervous system or other vital organs. A further incident was quoted by Ratcliffe (1954). A saki (*Pithecia monacha*) died ten days after it arrived at the Philadelphia Zoo; a uakari (*Cacajao rubicundus*) and a squirrel monkey, which had been exposed to the saki, also died of the disease but their mates were not affected.

Old-World monkeys and apes have, on the whole, proved refractory to experimental infection with *Toxoplasma gondii.* It must, therefore, be assumed that, whereas apes and Old-World monkeys are relatively safe, captive New-World species represent a danger to man, and particularly to children who develop more severe symptoms than adults. The mode of transmission is unknown, but experience at the Philadelphia Zoo indicates that the danger from pet South American monkeys is very real. Sureau *et al.* (1962) have also described the disease in a pet lemur in Madagascar, and – in spite of the failure to infect Old-World monkeys experimentally – Pokorny *et al.* (1961) isolated the parasite from *Macaca irus.*

In conclusion; *Toxoplasma* sp. has been found in a number of monkeys by various workers. It is only slightly – if at all – pathogenic, except in South American species.

7 Fungi

Fungal diseases are usually classified as:

1 superficial mycoses,
2 deep mycoses.

The superficial mycoses are caused by the related groups of fungi, known as the dermatophytes, which cause various forms of ringworm and favus.

The dermatophytes have the unusual property of being keratolytic; that is they can dissolve keratin in the horny layers of the skin. They are solely parasites of the skin and hair and are further sub-divided into (a) those which affect hairless parts of the body, belonging to the *Trichophyton* group, (b) those which attack hairy regions, belonging to the *Microsporum* group. There is another group of skin fungi, which cause diseases known as black and white piedra. These are fungi known as *Piedraia* and *Trichosporon*, and are in some ways intermediate between the two foregoing groups. *Trichophyton* solely parasitises the skin; and *Microsporum* is found within the hair and hair follicles; the *Piedra* groups invade the skin, and also form colonies on the surface of and inside the hairs.

The deep mycoses are caused by a heterogeneous collection of fungi, the majority of which are budding yeast-like individuals; but some are mycelial like the pathogenic mould fungi, e.g. *Aspergillus*. Whereas the dermatophytes are readily transmitted from one host to another by simple contact, the deep mycoses are mostly acquired from an infected habitat, usually from infected food or hay. Thus animals may be reservoirs, and the diseases may be zoonoses in the sense that reservoir animals contaminate their habitats, and man picks up the infection by eating food or inhaling spores from the contaminated soil or vegetation. The animals mostly incriminated as reservoir hosts of the deep mycoses are rodents, because the infecting fungi are saprophytic in soil or vegetation used in nest lining. The rodents themselves live so constantly in contact with these fungi that in endemic areas they are usually symptomless carriers, and thus spread the infection far and wide. Primates could well be reservoir hosts for some fungal species, such as *Histoplasma*, and their habit of regularly sleeping in the same trees night after night might place them in a similar position in an arboreal cycle to that occupied by rodents in a terrestrial one. The position of the primates remains a possibility only, as there is no knowledge or evidence to clarify the subject.

Furthermore, it is not known to what extent either superficial or deep mycoses are endemic diseases in the different primate groups. No genera or species of fungi are definitely associated with primates, although they can be readily infected with most of those species which can infect man. Reports of mycoses in primates are rather scanty, and there are still fewer which indicate with any degree of certainty that the infection was acquired by the primates in the wild state. These reports, however, must be regarded with some reserve, in view of the findings of Kaplan (1959) of a very high incidence of black piedra in primate pelts imported from Asia, Africa and South America (see below).

In captivity, monkeys can acquire ringworm or other fungal infections either from man or from dogs and can then transmit the disease to other primates including man. Monkeys can be infected by dermatophytes, acquired from human sources, but when so infected they do not readily infect their companions; but they can very easily transmit ringworm to human beings.

Many of the deep mycoses, e.g. the candidiases, originate from fungi which may be commensals in either man or simian primates, and which only cause overt symptoms of disease when the host is in ill-health or under stress. It is possible that such mycoses may become of greater importance, because both in man and simian primates they tend to compete with the normal bacterial flora, becoming dominant if this is destroyed by antibiotics. Thus mycotic disease may often be a sequel to normal antibiotic therapy, and in this case no external source of infection will be found.

It is sometimes said that the deep mycoses cannot be true zoonoses, because they cannot be transmitted by direct animal-to-animal contact or through an intermediate host, but only in a contaminated habitat. On the basis of the definition of zoonoses given in the first chapter, this reasoning would appear to be unsound, and it will be necessary to consider which fungi could be mutually infective to human and simian primates, even though the contact between the two groups is at second-hand by way of the habitat, or even acquired at third-hand via rodent and habitat, giving a cycle of this type: monkey–habitat–rodent–habitat–man.

Dermatophytoses (Superficial mycoses)

Pinoy (1912) first described a mycotic, cutaneous disease in monkeys, discovered by Levaditi. The infection was transmissible from one animal to another. Lesions were found mostly in the relatively hairless parts of the body on face and abdomen. The fungus was cultured and inoculated into guinea-pigs. It was thought by Pinoy to belong to the genus *Epidermophyton*, which is a group otherwise entirely confined to man, and he named it *E. simiae*. Chalmers and Castellani (1919), however, in the third edition of their *Manual of Tropical Medicine*, described it as *Pinoyella simiae*, preferring to give it a genus of its own. Dodge (1935) in his *Medical Mycology* retained Castellani and Chalmer's name for this fungus. The same fungus was isolated from a monkey by Emmons (1940), and was studied by him. This author placed it in the ubiquitous *Trichophyton mentagrophytes* group; this classification has since been accepted without question. *T. mentagrophytes* commonly affects dogs and cats and also man. It also affects most groups of rodents.

Weidman (1935), when describing the dermatoses of monkeys and apes at the IXth International Congress of Dermatology at Budapest, quoted two cases of dermatophytoses in Barbary apes (*Macaca sylvanus*). Both were affected with extensive rings and hairless patches of white scales on abdomen, head and the anterior surface of the arm. Mycelium was present in the sheaths of the hair follicles, and *T. gypseum* was isolated. The same author (1927) previously described ringworm in a gorilla and two chimpanzees, associated with infection by *T. equinum*. This is a ringworm fungus found in horses, which has not been described so far as a saprophyte. How these apes became infected was not known.

Voronin *et al.* (1948) described an extensive epizootic of ringworm in a group of baboons in a single cage at Sukhumi. Sixty animals were infected some having as many as thirty separate lesions. The cage occupied by these baboons had housed infected dogs a year previously, and had evidently not been sufficiently cleansed and sterilised. The organism involved was identified as *T. violaceum.*

It is noteworthy that in the very few cases described of trichophytosis in simians, there is absolutely no suggestion that infection had been present before the animals were brought into captivity; conversely it would seem probable that all cases were derived from contact with man or domestic animals.

Fox (1923) in his *Textbook on Diseases in Captive Wild Animals and Birds* described no cases of ringworm, but (1926) he described a ringworm widely disseminated over the more or less hairless areas of a young chimpanzee. Scrapings of skin contained abundant mycelium, but cultures were unsuccessful. The distribution of lesions reminded Fox of *Tinea cruris* in man, and salicylic ointment produced a marked improvement. The animal's cage-mate was strangely enough not affected. *T. cruris* in man may be associated with a variety of dermatophytes, belonging to the *Epidermophyton* and *Trichophyton* groups.

Fungi which affect hairy areas, attack the hairs, the horny linings of the hair follicles or even the nails. Such organisms affecting man in the USA are *Microsporum audouini* and *M. canis*; *Trichophyton schoenleini*, *T. violaceum*, *T. tonsurans*, and *T. sulfureum* and those resident in some European countries are also present. These fungi vary in their depth of penetration and are relatively host-specific; those highly infectious for animals being less easily transmitted between men. Weidman (1923) attempted without success to transfer to monkeys a variety of these fungi infecting man. In general, monkeys are not especially susceptible to the human fungi, but they can sometimes be infected and can then pass human ringworm back to man.

Kaplan *et al.* (1957, 1958) studied ringworm in fourteen New-World monkeys infected with *Microsporum canis* or *M. distortum*. The commonest sites for lesions were the extremities and the tail; but in four *M. distortum* cases, scaliness and baldness spread over the entire body from a lesion on the head. Contact cases occurred in human beings from five of ten of the *M. canis* cases and three of four of the *M. distortum* cases; these were typical ringworm. The *M. distortum* cases are the only ones so far described in man in the USA. Three of the four monkeys were three-to-four-month-old *Cebus*, purchased at intervals from a single pet shop. The fourth was an *Ateles*, which was obtained from another dealer. The infections were probably imported with the monkeys from South America, but there is no knowing whether they were contracted in the wild, or after capture.

Records at the London Zoo contain reference to an outbreak of ringworm amongst Lar gibbons, which occurred during 1961 to 1963, and spread sporadically through the den. These gibbons appear to have been imported carrying the infection, since they were never in contact with other animals at Regent's Park. The fungus was identified as *M. canis*. It responded readily to treatment.

Kaplan and his fellow authors examined hair or skin scrapings from one chimpanzee and twenty-one New-World monkeys with cutaneous lesions; the specimens having been submitted from all parts of the United States. No fungi were found in the scrapings from the chimpanzee, but fifteen monkey specimens (71·4%) contained one or another of three species of fungi. The commonest (ten cases) was *M. canis*, presumably contracted from domestic dogs or cats. Four monkeys were infected with *M. distortum*, not previously reported in the USA, and therefore possibly a natural disease of the monkeys. One animal harboured *Trichophyton cutaneum*. Since this is very rare in the States, it too was probably imported with the monkey.

Mulzer and Nothaas (1926) inoculated *Macaca mulatta* with syphilitic material from humans; and amongst other lesions ringworm, described as a spreading microsporosis of the skin, was transmitted. Microsporosis of monkeys was also described by Ajello (1953), due to *Microsporum gypseum*, and by Conant (1937).

Lochte (1937) described five cases of black piedra in chimpanzees at a German zoo, caused by *Piedraia hortai.* Kaplan who, together with other workers (Kaplan *et al.* 1958), had found white piedra in a black spider monkey from South America, caused by *Trichosporon cutaneum*, later conducted a large-scale study on 438 primate pelts from museums, of which 195 (44·5%) were found to have black piedra. Of the pelts imported from Asia 60% showed this disorder, as did 28·7% of African primate pelts and 76·1% of New-World animal skins. By contrast, only 6% obtained from the primates kept in the New York Zoo were found to be infected. These findings establish the frequency of black piedra in the hairs of different primate species as well as in man and suggest the existence of a wild primate reservoir of infection in the forests of tropical and semi-tropical Africa, Asia and Latin America.

The great dearth of references to ringworm in simian primates indicates that the disease is neither common nor important, and that these animals are not a source of serious danger to human hosts. They can, however, quite readily be infected by dogs and, with less ease, by humans; after infection by either source they can transmit the disease to human hosts.

Deep Mycoses

The deep mycoses can be divided into two groups: first, those, generally mould fungi, in which mycelium is present in lesions; secondly, fungi which are present in lesions in yeast-like structures which may under certain conditions of culture produce mycelium. The fungi which produce mycelium are all acquired from infected vegetable material; thus as zoonoses their importance lies in the possibility that simian primates might form a reservoir of infection and thus contaminate human food crops. In the case of aspergillosis in monkeys, a peculiar symbiosis of fungus and *Mycobacterium* has been reported within the actual tubercles which thus carry a dual infection. The significance of this in relation both to aspergillosis and to tuberculosis as zoonoses will be considered later.

Apart from *Aspergillus*, reports of simian infections with mycelial fungi, also pathogenic in man, are scanty. A single case of mucormycosis caused by one of the common mould fungi, *Absidia* sp. or *Mucor* sp., was studied by Fiennes at Regent's Park (unpublished data). This case was alimentary, and the large bowel was peppered throughout its length with pinpoint, black, shining ulcers. In the rectum there were longitudinal haemorrhagic plaques where the ulcers had coalesced, the edges of these were raised and irregular; between the ulcers, the mucous membrane was necrotic and stripped easily. The patient was a variegated spider monkey (*Ateles belzebuth hybridus*), which suffered severely from rickets. A second case which was not investigated histologically, also in a spider monkey (*A. hybridus*), showed similar lesions supposed to be due to the same cause.

On one occasion, a pathogenic species of *Penicillium* was isolated from an ape at Regent's Park. The individual was a mountain gorilla, a juvenile male, which died from bronchopneumonia. *Pneumococcus* type III was isolated from the lungs, together with *Penicillium piceum*; the latter is usually regarded as pathogenic.

Sporotrichosis has not been described as a spontaneous infection in monkeys, but they are susceptible to it. This fungus grows on shrubs, and infection can occur from a minor thorn puncture. Benham and Kesten (1932) grew *Sporotrichum schenkii* from a human source on *Berberis* thorns. One of these spores implanted in the right index finger of a monkey resulted, after a few weeks, in generalised swellings of lymph nodes, and specially those of the right axilla. Additional organisms injected intravenously caused the development of papules and nodules, arranged linearly along the extensor surface of the limbs. The distribution was similar to that in the human case and suggested spread through lymphatic channels. Though not described as such, this infection could be a possible zoonosis, where monkeys and men use the same pathways through thick country containing thorny trees.

Aspergillosis, caused by *Aspergillus fumigatus* or *A. flavus*, is a common infection of the respiratory passages of all groups of animals. It is particularly severe and serious in birds. Infection occurs by the inhalation of spores from contaminated vegetable material. In mammals the fungus does not usually cause serious disease, although infections may become acute when the normal bacterial flora of the respiratory system is eliminated by antibiotic therapy. Amongst simian primates there appear to be no references to aspergillosis except those from the London Zoo. These infections occurred during a limited period of years, being reported by Scott (1926, 1927, 1928) and by Hamerton (1929). In most of these cases the fungus (*A. fumigatus*) was present in tubercles in company with *Mycobacterium tuberculosis*. Scott (1930) reviewed his cases in the Medical Research Council publication *Tuberculosis in Man and Lower Animals*. It was Scott's opinion that the mycotic infection preceded the tuberculous and that the acid-fast bacilli showed a predilection for invasion of mycotic nodules. As is usual with Old-World monkeys, tuberculosis once established proved rapidly fatal. It is evident that both man and his simian cousins can contract aspergillosis from a common source, and this could contribute to the establishment of tuberculosis in an otherwise susceptible subject. Scott recorded the combined disease in two species of lemurs (*Lemur varius* and *L. catta*), the pig-tailed macaque (*Macaca nemestrinus*), the Bonnet macaque (*M. sinicus*), the Wanderoo (or lion-tailed) macaque (*M. silenus*), the Rhesus macaque (*M. mulatta*) and the vervet monkey (*Cercopithecus aethiops pygerythrus*).

The organs involved in the combined infection were the lungs, liver, spleen, kidneys and mesenteric glands. Hamerton (1929) described a case in a Roloway monkey (*C. diana*) as follows: 'The infection was mostly mycotic and was alimentary in origin; the primary foci were present in the mesenteric glands. Liver and spleen were thickly studded with nodules of mycotic growth, in which acid-fast bacilli were sparsely scattered. The animal was apparently well-nourished.' The bacilli were described as morphologically resembling *Mycobacterium tuberculosis*. Hamerton suggests that there is symbiosis between fungus and bacillus and draws the analogy of human leprosy, where *Streptothrix leproides* is associated with *Bacillus lepri* in leprous nodules.

Scott (1930) in his review, mentioned above, states that when tuberculosis and mycosis are found in association in man it is usually as a super-infection or late implantation of the latter on the former; the same is believed to be true for birds and domesticated animals. This order of events appears to be reversed in monkeys, probably because the mycotic infection is nearly always chronic, whereas tuberculosis runs an acute course, and death occurs from it before the mycosis, if present, has had time to develop. The accounts given both by Scott and Hamerton generally stress the severity of mycotic lesions and the relative mildness of the tuberculosis. This might suggest that the presence of mycosis modifies the usual, galloping nature of tuberculosis in Old-World primates. Scott makes the further point that all the monkeys found infected with both tuberculosis and mycosis were Old-World species which are much more susceptible to tuberculosis than the New-World monkeys.

Since the reports by Scott and Hamerton, aspergillosis has only been diagnosed once in a primate at the London Zoo, from a nasal swab taken from a mountain gorilla. This is, of course, doubtful evidence of a respiratory infection. It is evident from what has been said, however, that *post-mortem* studies on animals known to have died from respiratory infections should include cultural and other investigations to see whether aspergillosis was also present. It is probable that mixed infections have been frequently overlooked.

Amongst the yeast-like fungi, Catanei (1925) appears to have been the first to describe the presence of *Candida albicans* as a benign mycotic infection in two of twelve Algerian monkeys (*Macaca sylvanus*). Tongue scrapings revealed the presence of round or ovoid, yeast-like organisms 4–6 μ in diameter, but there were no noticeable lesions, and the tongues of the affected monkeys appeared quite normal, the mucous membranes being the same colour in infected and uninfected animals. *C. albicans* has also been isolated (unpublished data) by the author at the London Zoo, both in circumstances where it appeared to be commensal, and also where slight or severe lesions were present resembling 'thrush'. It has been isolated from woolly monkey (*Lagothrix* sp.), capuchin (*Cebus* sp.) and gorilla (*Gorilla gorilla beringei*), particularly in cases where lesions were present. In gorilla and woolly monkey, *Candida* infection was limited to the tongue, but in the capuchin it was isolated from the tongue, the roof of the mouth, lungs, liver and intestines. This case was a generalised mycosis, and death of the animal appeared to have been due to this cause.

Young *et al.* (1957), in an article *Care and Management of a Macaca Mulatta Monkey Colony*, made the point that candidiasis in these animals is usually secondary to other infections, such as intestinal parasitism or dietary deficiency, or as a complication of antibiotic therapy. They state that the site of attack may be the skin, lungs, intestinal tract, reproductive tract or nervous system. The signs vary accordingly. Pet monkeys infected with *C. albicans* could thus be a danger to children. There are, however, no reports that such infection has ever occurred. Reimann and Kurotchkin (1931) failed to transmit *C. tropicalis* from an infected human to *Macaca sinica*, although mycosis was established in rabbits, guinea-pigs and mice. Even traumatising the lungs of the monkeys did not render them susceptible to infection.

The reaction of Old-World monkeys to injection with *Torula* and other yeasts and yeast-like fungi seems to vary considerably and no indisputably spontaneous infections have been reported. South American monkeys, on the other hand, are more uniformly susceptible and generally more severely affected. Mycotic infections (deep mycoses) have long been recognised in monkeys held in zoos, and some are probably attributable, as with coccidioidomycosis, to spores present in straw or hay (inhalation). Many of these infections, at necropsy, resemble generalised tuberculosis, but tubercles must be cultured for *Pasteurella* and fungi, as well as for tuberculosis. Takos and Elton (1953) described spontaneous cryptococcosis (torulosis) due to *Cryptococcus neoformans* in marmosets (*Callithrix geoffroyi*) which were captured and kept in Panama. The marmosets were found dead after three months in captivity. The lungs were deep purplish black and congested, with a rubbery consolidation throughout. All chambers of the heart were markedly dilated. The spleen was enlarged and ovoid, with soft, jelly-like consistency. The mesenteric lymph nodes were greatly enlarged (3–15 mm) and pale grey. Most alveoli of the lungs were filled with an exudate of plasma cells and lymphocytes, and some contained blood. There were birefringent bodies (3–8 μ in diameter). Many budding cells were found. The lung lesions were similar to those of human torulosis cases, differing only in intensity and wide dissemination. The authors considered that the disease in marmosets was liable to take a more explosive course.

Cryptococcus neoformans has been isolated twice from simian primates at the London Zoo: first, from a slender loris (*Loris tardigradus*), in which the skin and hair were infected but there was no generalised disease; secondly, from a Brazza's monkey (*Cercopithecus neglectus*), which apparently died from a condition of shock, in this case there was a great thickening of the stomach wall in the region of the pylorus, and *Cryptococcus neoformans* was cultured from the tissues.

Weidman (1933) showed that *C. neoformans* could be transmitted from man to monkey and he transmitted the infection from a human case. Weidman's monkeys produced lesions at the site of injection, but the infection did not become generalised. Kessel and Holtzwart (1935) also injected *C. neoformans* from a *torula* infection of the knee in a man. The material was introduced to *Macaca mulatta* by intracardial injection. Subcutaneous nodules, elevating the skin, appeared in about a month. One monkey recovered; in the other, the infection spread to the brain, but not to the viscera. In a subsequent series of experiments with a strain of human torulosis, derived from the nervous system, no cutaneous lesions were produced. There were, on the other hand, lesions both on the surface of the brain and within the cerebral cortex. Smears from the brain, the spinal cord and the viscera contained fungi.

4+

London Zoo, *Coccidioides immitis* being cultured from mycotic lesions in the lungs of a spider monkey (*Ateles hybridus*). The latter case is possibly more plausible than Hamerton's, because at least the monkey involved was imported from South America.

The most celebrated case of coccidioidomycosis in a simian primate was that of the mountain gorilla (*Gorilla gorilla beringei*) at San Diego Zoo described by McKenney *et al.* (1944). These authors also described a case in a capuchin (*Cebus hypoleucus*). San Diego Zoo is in the centre of the endemic area for coccidioidomycosis, and there is continual danger of the introduction of the disease to primates and other susceptible species. The gorilla possibly became infected from contact with infected hay, but some authorities believe that spores can be carried in dust blown by the wind. Infection appears to take place chiefly by inhalation. This gorilla lost interest in food and became less active. On the thirty-second day it appeared to have fever. Dyspnoea developed on the forty-first day, and there was intermittent haemorrhage from the right naris. Meanwhile there was no evidence of pain and no persistent cough. It died on the forty-fifth day. Autopsy revealed eight soft white areas, 0·5 cm in diameter, in the liver. The spleen was soft and pulpy, containing two lesions, 1 cm in diameter and at least twenty-five smaller lesions like soft tubercles. The right pleural cavity contained about 1 litre of brownish-green fluid. On the pleural surface of the diaphragm, there were haemorrhagic excrescences 1–2 cm in diameter, and numerous tubercles were scattered over both lungs. The central zone of the lower lobe of the lung was occupied by an irregular abscess cavity filled with putrid material, and surrounded by 'beefy' tissue, resembling a neoplasm of granulation tissue. There were no cutaneous or subcutaneous granulomata. Fungi were present in large numbers in liver, spleen and lung. Coccidioidomycosis was proved by cultural and immunological tests.

Emmons (1942) isolated *Coccidioides immitis* from soils in Southern California and was able to demonstrate (1943) that there was almost universal infection of a silent nature in wild rodents. He gave a further account (1947) of the biology of *Coccidioides*.

The aetiological picture of this strange fungus is – distribution apart – perhaps typical of what may be expected with fungi of this kind. The fungus can survive, if not multiply, in the sandy soils of this region and it forms a reservoir amongst wild rodents, being especially common in their nests. It was found by Emmons in 15% of 124 pocket mice (*Perognathus baileyi, P. pennicilatus, P. intermitis*), and 17% of 29 rats (kangaroo rats, *Dipodomys merriami*). The fungus was not found in 113 captured specimens of *Peromyscus eremicus*. Both humans and simians can be readily infected with the disease from soil by a variety of means. A pet monkey infected with this fungus could be a dangerous companion, either for adults or for children.

Reports of actinomycosis in simian primates appear to be limited to those of Weidman (1927, 1928 and 1935), who described two cases in capuchins (*Cebus capucinus*). He described the presence of ray fungi together with closely branching gram-positive threads demonstrated in the discharge from an abscess along the side of the nose. The condition was successfully treated by ultraviolet therapy. The same condition was also observed in *Ateles geoffroyi* and similarly treated.

Possibly actinomycosis is less rare than might be inferred from this one reference. Records at the London Zoo (author, unpublished data) show diagnosis of the disease in two simian primates, as follows:

One, a moor monkey (*Macaca maura*), an adult male which had been in the Zoological Gardens at Regent's Park for twenty months, died as a result of acute enteritis. The gut was greatly distended and there was acute haemorrhagic enteritis of the small intestine, in which the mucous membrane was extensively sloughed. The contents of the large bowel were fluid, and the mucous membrane was inflamed. Bacteriological cultures revealed the presence of a heavy concentration of *Actinomyces* sp. in the gut wall and contents.

The second, a young male gibbon (*Hylobates moloch*) died in 1963, after five months in the Regent's Park Zoo, and a diagnosis of generalised actinomycosis was made. The liver at autopsy was found to be studded throughout the substance with caseous nodes, the largest being some 2 cm in diameter. The pancreas was caseous, but lungs, kidneys and intestinal tract appeared normal. Bacteriological culture revealed the presence of *Actinomyces* sp., probably *A. israeli*. It would appear, therefore, that, though rare, actinomycosis may sometimes occur in non-human primates and, owing to the nature of the disease, there could be some danger of direct cross-infection to man.

Apart from the cases given above, de Monbreun (1935) studied a disease in monkeys, caused by *Monosporium tulanese*, which had been isolated from a verrucuous lesion on the face of a patient. Cultures of yeast-like organisms were injected intradermally in three *Macaca mulatta*, and after four weeks areas of induration appeared at the injection site. These areas progressed to flattened nodules confined to the skin, and eventually became ulcerations about 1 cm in diameter. Subcutaneous injection caused deeper pathological changes.

This last case perhaps is illustrative of the situation in regard to most fungi as primate zoonoses. Simian primates do not appear to be naturally affected by fungi to nearly so great an extent as are human primates, but they can mostly be infected with those that are pathogenic to man. The infections caused in this way are often progressive and fatal, and it may be rather an accident of geographical habitat that has enabled members of the monkey kingdom to avoid most of these unpleasant conditions. Nevertheless, the fact remains that man's primate cousins can become infected and do become so infected, even if rarely, under conditions of captivity. Any monkey, therefore, purchased from a dealer as a pet could transmit any of the human dermatomycoses by direct contact and most of the other pathogenic fungi at second-hand, by contaminating lawns or sandpits where children play. It is thus important, that these diseases should not be overlooked when monkeys are quarantined and screened before being accepted as fit house-mates.

Whether these dangers arise more from the transmission of infections caught before capture, or acquired from the animals' captors, cannot be determined.

Part III Bacteria

8 Bacteria

Neither apes nor monkeys have so far been suggested as reservoirs of bacterial diseases to which man is susceptible, with the possible exception of relapsing fever in tropical Africa, and recurrent fever in South America. On theoretical grounds, there would appear to be few dangers of bacterial disease spreading from man to other primates or from other primates to man, under wild conditions. Such cross-infection could occur in the case of diseases to which both were susceptible and which could be spread by means of contamination of water supplies, or by other animals such as wild rodents, armadillos or marsupials such as opossums. Those diseases which could be spread by contamination of water supplies would be cholera or salmonellosis, and by other feral animals, salmonellosis, pseudotuberculosis and actinomycosis.

It has not been shown that this ever happens, and the importance of simians in the spread of disease to man lies mainly with captive apes or monkeys which are in direct contact with human beings. In this sense, both apes and monkeys are potentially a source of danger to their captors, because they are often symptomless carriers of a number of organisms which can cause serious diseases in man and particularly in children. The main danger arises from the dysentery sub-group of organisms, which are often present in the faeces of apes and monkeys, particularly gibbons, although the animals show no clinical signs. Apes and monkeys are also susceptible to a wide range of organisms, which affect human beings. Captive simians, however, are more likely to acquire infections from human beings than human beings from simians because of the restrictions placed on them. These points will be amply illustrated by a survey of the bacterial organisms associated to the greatest extent with disease in apes and monkeys, and of the extent to which they are believed to pass between simians and human beings. The spirochaetes will be discussed first because of the possibility that relapsing and recurrent fevers may be zoonoses under feral conditions.

Spirochaeta

The important human disease of relapsing fever, present in almost all continents of the globe, is caused by different species of *Spirochaeta* (*Borrelia*). In human communities, it is associated with dirty conditions, and is transmitted by lice or bed bugs. Transmission is by contamination of the wound caused by the bug or tick with its faeces. Tictin (1897) working in Russia found that he could infect monkeys by inoculation with the blood obtained by crushing bugs which had very recently fed on a relapsing-fever patient. Earlier than this Vandyke Carter (1877), who discovered the cause of relapsing fever to be a spirochaete, not only inoculated himself on two separate occasions at an interval of two-and-a-quarter years, but also successfully reproduced the disease in monkeys by the injection of his spirochaetes. Both in Africa and South America, spirochaetes identical with those of man, *S. recurrentis* in Africa and *S. neotropicalis* in South America, are natural parasites of simian primates living under feral conditions. In neither continent does it appear that the spirochaetes do more than cause mild fevers, except possibly in very young animals. In older animals there appears to be a developing immunity, which corroborates the view that spirochaetal infection is a naturally occurring endemic disease of wild primates. As a zoonosis, the dangers of which arise from captive simians, the disease is plainly of little significance. This spirochaetosis may, however, well be an important zoonosis under feral conditions and simian primates important reservoir hosts.

Literature, referring to natural and experimental spirochaetosis in simian primates, was reviewed by Nouvel (1954). Only a few naturally occurring infections had been described in the great apes and Old-World monkey species, but most genera, including

Cercopithecus, *Macaca* and *Papio*, are readily susceptible to experimental infections. Amongst South American monkeys, spirochaetosis is quite common and they are thought to form, together with rodents, an animal reservoir from which outbreaks in man originate. If this is true of South America, one may suppose it also possibly true in Africa and Asia. The important point to consider in relation to these areas is whether an eradication policy of the man–man disease would be defeated by the reintroduction of the disease from the co-existing arboreal cycle. It is extremely probable that this might happen.

The man–man cycle is maintained in most areas by body lice or bed bugs. These may perhaps be regarded as unusual, or unnatural, intermediate hosts. The more usual hosts are ticks of the Order Argasidae, and genus *Ornithodorus*. In many regions, for instance Western Uganda of which the author has first-hand knowledge, these ticks are the intermediate hosts in man as in primates. This Order of ticks is, however, best known as parasitic on birds. Characteristically, these ticks lodge in cracks or crevices, only emerging at night to suck the blood of the host. *Ornithodorus moubata*, which is the intermediate host of *Spirochaeta recurrentis* in man in Western Uganda, lodges in cracks in the bush poles of which the local people make their huts. Whereas the local people reach adolescence with a satisfactory degree of immunity, it is dangerous for immigrants to sleep in their huts and under certain conditions relapsing fever can become a major public health problem. For instance, again in Western Uganda, many immigrant labourers used to pass through from the neighbouring territory of Ruanda, and a high proportion contracted relapsing fever by sleeping in rest-huts provided on their route.

The natural history of this disease, with its unnatural intermediate hosts in the human subject, suggests the possibility that the natural reservoir may exist amongst wild animals, possibly including monkeys. Again, the natural vector, a tick which as its name implies is primarily a bird parasite, suggests the possibility that this disease may be derived from an arboreal cycle in which birds and simian primates are involved. It is symptomatic of our ignorance of the external parasites involved in arboreal disease cycles that we have no knowledge of whether *Ornithodorus* ticks do in fact prey on both birds and primates at upper-canopy levels, or indeed whether – as they surely must – they exist in these situations. We do know that similar spirochaetes (*Borrelia* sp.) are very common parasites in wild birds. For an account of the bionomics of these ticks, see pp. 39–41 (*supra*).

Research on this disease, as a possible zoonosis, would appear to be overdue. The role of simian primates in maintaining a reservoir may be important when eradication efforts are made, and transmission cycles could occur through wild rodents or birds. It is not easy to study these spirochaetes, since they cannot be maintained in culture and are difficult to isolate. They can only be transmitted from one host to another, under experimental conditions, through the agency of intermediate hosts such as argasid ticks. From such work as has been done it is not even known how many species exist, nor what is the infectivity to man of those so far isolated. This can only be determined by cross-immunity experiments, so their importance as zoonoses is virtually unknown. It may be suggested that they do, in many areas, pose important problems.

Both monkeys and man are susceptible to 'ratbite fever', caused by *Spirillum minus*. This is a natural parasite of rats and mice and is transmitted by bites from these rodents. Theoretically this disease could be transmitted to man by a bite from a monkey which had acquired the infection whilst in captivity from a rat or mouse bite, but there is no evidence that this has ever occurred.

Simians are equally susceptible with man to leptospirosis (Weil's disease) caused by *Leptospira icterohaemorrhagiae*. Ana-n'in and Karaseva (1955) tested eighty-five sera from hamadryas baboons and other monkeys for *Leptospira* antibodies; of these fourteen were positive. Ten of these came from forty-five hamadryas baboons and only four from forty other monkeys. One case of icterus apparently of leptospiral origin was diagnosed in a hamadryas baboon. Kavtardze (1958) artificially infected nine hamadryas baboons, five green monkeys and three rhesus. The only clinical cases resulting were in two four-to-six-months-old hamadryas baboons. All the other experimental baboons and monkeys developed inapparent disease, in which antibody titres rose to 1:10,000 and 1:20,000. Nouvel *et al.* (1957) also described outbreaks of leptospiral jaundice at the Parc Zoologique du Bois de Vincennes. The disease could be caused in captive monkeys by rat bites or via contaminated water or food and could possibly be transmitted in this way from monkey to man. Possibly some danger exists in this respect, because there is more likely to be contact between rats and monkeys in zoos than between rats and man; the monkey could thus act as an intermediate transmitter between rats and humans.

Amongst the treponematoses, various species of monkeys were used in early days for experimental work on syphilis. Although they are readily infected, there is plainly no necessity to consider this disease as a potential zoonosis. In his review of the treponematoses, Cockburn (1963) states that if man inherited his treponemes vertically from his remote primate ancestors, then a survey of the Order Primates would show that apes, monkeys and other members were also hosts to species of *Treponema*. These would not necessarily be pathogens. Indeed the origin of the treponematoses, including syphilis, yaws and pinta, is as yet a medical mystery. It is of interest that Cockburn's views are vindicated by the work of Fribourg-Blanc *et al.* (1963), in their immunological studies on what they call *Cynocéphale africain*, presumably Guinea baboons (*Papio papio*). They have come to the conclusion that endemic human yaws is probably maintained by a reservoir in these animals. This belief is based on positive antigenic reactions to treponemata found in African primate species. The authors used the immuno-fluorescence technique to detect serum antibodies and antigens, and to identify microbial strains. This technique allowed them to demonstrate an antigenic relationship between human and simian gamma-globulins. The animals tested with Nichol's strain of *Treponema pallidum* did not permit *T. pallidum* to be differentiated from *T. pertenue*, the agent of yaws, and from *T. carateum*. In 50% of the cases all species probably react to yaws antigen, and this fact suggests that eradication of endemic human yaws might also involve action against monkeys.

In previous studies, the authors had already come to the conclusion that it might be possible for wild-living simians to be reservoir animals for a virus infection dangerous to man, as quoted here (translated from the French original):

> May we recall at this point our observations made two years ago during the course of a very extensive biological study of 300 sera of Guyana Indians of absolutely pure race who lived in complete isolation in the Amazon forest: We found then that only 7 cases showed any unmistakable treponema antibodies, in a region where syphilis is unknown and other treponematoses endemic. We were, at the time, rather worried by this very low endemic index in a population known to possess a very high natural immunity but to be living in deplorable unhygienic conditions. Perhaps it is possible now to envisage the Amazon monkeys as being responsible for these sporadic cases.

Dr C. J. Hackett (1966, personal communication), a well-known authority on treponemata, confirmed that the tests used were recognised as entirely specific to treponemata, and he has no doubt that the animals' positive reactions showed that they had, at some time, been infected with a treponema. He also stated that treponemata have since been isolated from simians.

Further details of this work will be awaited with interest. Meanwhile the treponematoses, especially yaws, may well emerge as amongst the more important zoonoses.

Infections caused by Cocci and other miscellaneous organisms

Streptococci, staphylococci, meningococci and pneumococci all occur as normal inhabitants in the nasal passages of monkeys and apes. All may, under certain conditions, cause serious diseases which could infect human beings. The commonest organisms associated with pneumonia in apes and monkeys, but especially in chimpanzees, are the pneumococci; these may be of any of the four main types. Unlike the situation in humans, investigations at the London Zoological Gardens have failed to establish that one type is more commonly found in association with pneumonia than another, and any type may be found as a normal inhabitant of the respiratory passages. Cases of pneumonia frequently appear in young chimpanzees, either when they are newly imported, or when they are subjected to adverse conditions such as chilling at the beginning of winter. The chimpanzees usually make a good recovery without subsequent relapse, if promptly treated with drugs to which the organism is sensitive. The condition is more serious in gorillas and gibbons which frequently die. Pneumonia in chimpanzees associated with pneumococci often appears in epidemic form, but seems to be quite unrelated to epidemics in man, and the danger that an epidemic amongst chimpanzees could start one amongst their human keepers appears slight.

Staphylococcus albus and *S. aureus* have been found in association with pneumonia and septicaemic conditions in monkeys, and the same can be said of various strains of streptococci. As with the pneumococci these are potentially pathogenic to man, but no cases have been described in which human disease has been transmitted from simian patients. Obviously, however, simians showing signs of respiratory disease must be regarded as potentially dangerous, especially to children.

Rare cases have been recorded in London of meningitis associated with meningococci, and one case with *Micrococcus tetragenes* with fatal results. The path of infection appeared to be from the respiratory passages by way of infected sinuses and so to the brain. Here again, simians could be potentially dangerous and the case for care with children is reinforced.

Lapin and Yakovleva, in their *Comparative Pathology in Monkeys* (1963, but originally published in Russian 1960) have reviewed the causes of death amongst large numbers of monkeys of different ages and groups, kept at the Sukhumi Primate Centre. Pneumonia was the main disease in the overwhelming majority of cases involving all species of monkeys over one month old. They considered the results of investigating 117 cases of pneumonia, autopsied in the Institute between 1952 and 1958. The cases were divided into two groups, chiefly according to morphological features and classified as bronchopneumonia or lobar fibrinous pneumonia. Most of the cases of bronchopneumonia involved the macaques, there being sixty-three macaques, twenty-four baboons and fourteen green monkeys amongst the 101 bronchopneumonia cases. Bronchopneumonia generally ended in recovery with resorption of the exudate. Only sixteen of the 117 pneumonia cases were of the lobar fibrinous type. This is essentially pleuropneumonia, and there are very thick films of fibrin enveloping the lung surface. Lobar pneumonia is often associated with a high death-rate, unlike bronchopneumonia. The authors state definitely that lobar pneumonia in monkeys is associated with pneumococci, but they were unable to make bacteriological investigations, and the pneumococcal types were not identified. The cause of bronchopneumonia in their cases is not stated, although one case is cited where the pneumonia was followed by pneumococcal meningitis.

Lund and Petersen (1959) stated that pneumococcal infection was widespread in their monkey colony in Sweden. They considered, however, that baboons were less susceptible to pneumococcal pneumonia than other monkeys. Amongst 151 autopsies, almost 50% had pneumonia, but none of the four baboons had the disease. It is possible, of course, that the lack of pneumonia deaths amongst baboons merely showed that these animals recover better from infection.

Ivanov and Aksenova (1961) investigated 278 cases of pneumonia in macaques, hamadryas and green monkeys. Bronchopneumonia was present in 74% of the cases, 'membranous pneumonia' in 26%. The causative agent in both types of pneumonia is stated to be *Pneumococcus*, but the species were not given. In cases of lobar pneumonia, Friedlander's *Pneumobacillus* is sometimes the cause.

Respiratory disease in monkeys has also been reviewed by Sauer and Fegley (1960). Bacterial pneumonia is stated to be the most serious respiratory disease encountered in monkeys, and bacteriological investigation of thirty-two infected lungs showed that the following organisms were involved: *Pneumococcus* spp. (13), Beta-haemolytic *Streptococcus* (11), *Staphylococcus* sp. (9), Alpha-haemolytic *Streptococcus* (1) and *Klebsiella pneumoniae* (Friedlander's *Pneumobacillus*) (1).

These authors also observed giant-cell pneumonia in a *Macaca mulatta* and in a *Cercopithecus aethiops sabaeus*. In these two cases the microscopic lesions were those of an interstitial pneumonitis with large, multi-nucleated, foreign-body-type giant cells. Eosinophilic inclusion bodies were seen within nuclei of epithelial cells of the bronchial mucosa,but not within the giant cells. Cytoplasmic inclusion bodies were present, though not prominent.

Vanace (1960) reported attempts to produce the counterpart of rheumatic fever in the rhesus monkey (*M. mulatta*) by means of pharyngeal infection with group A beta-haemolytic streptococci. Results of cardiac tissue studies demonstrated no quantitative or qualitative differences from the control normal monkeys, and no Aschoff's bodies were produced. Of the monkeys used, one sustained an active, and six developed sub-clinical infections with streptococci. Seven monkeys demonstrated a rise of titre to the test streptococci.

To what extent simian pneumococci, whether from carrier cases or from diseased animals, represent a hazard to human health cannot be stated. It is certain that most groups of monkeys do carry all groups of pneumococci, and under certain conditions they do transmit them to each other in pathogenic and epidemic form. It is sensible for children at least to be protected from monkeys showing signs of respiratory disease, although cases of disease amongst the keepers of infected animals do not appear to occur very often.

Klebsiella pneumoniae (Friedlander's *Pneumobacillus*) is commonly found as a commensal in the mouth of monkeys and apes. It is sometimes also associated with conditions of pneumonia and could possibly be dangerous to humans, especially children.

Amongst the diphtheroid bacilli, diphtheria is not a natural disease of apes and monkeys, and no cases have been recorded. Diphtheroid organisms are commonly found, however, as normal inhabitants of mouth and gums, but are not apparently associated with disease. *Corynebacterium pyogenes* has been incriminated in septicaemic diseases in a gorilla, chimpanzees and woolly monkeys in London.

Amongst the pasteurelloses, neither monkeys nor apes have been suggested as reservoirs of plague either in Africa or India.

Pasteurella multocida is occasionally found as another mouth commensal in simians, but there is no evidence as to its association with disease. From 1962 to the present day, occasional cases of pseudotuberculosis, caused by *P. pseudotuberculosis*, have occurred at the London Zoological Gardens. Among the simians, a Schmidt's white-nosed monkey (*Cercopithecus ascanius schmidti*), another of the cercopitheques (*C. nictitans signatus*), and one of the apes, a black-and-grey gibbon (*Hylobates moloch funereus*) died of the infection. This trouble was heralded by cases amongst rodents, such as agoutis and coypus, and some birds. It is probable that the infection was introduced and is maintained in wild birds, whose droppings contaminate the food of the captive fauna.* Occasionally, this organism infects children, in whom enlargement of the mesenteric glands, associated with abdominal pain, may occur. There is a rather remote possibility that a pet monkey, which was allowed in the garden and picked up food, possibly contaminated by bird or rodent faeces, could contract the infection and pass it on to the child. This, however, is perhaps no more of a risk than the danger from infection by pet guinea-pigs, which among the rodents are most likely to be closely associated with children and have also been found to be one of the mammals most frequently infected with *P. pseudotuberculosis* (Mair, 1965).

Pseudomonas aeruginosa is occasionally found in the faeces of monkeys, and at least one monkey has died at the London Zoo as a result of generalised infection with this organism; this was manifested by pyaemic abscesses in the internal organs, particularly the spleen. The death of a female orang-utan (*Pongo pygmaeus*) which occurred at the Zoological Gardens in November 1965, was also attributed to *Pseudomonas aeruginosa*, which had caused septicaemia and toxaemia. Both kidneys were affected with pyelo-nephritis, the lungs were very oedematous, and the alimentary tract showed a longitudinal ridge of haemorrhagic tissue running along its length, which was due to sloughing mucous membranes. No parasites or fungi were found in the

* *Vide* also Nouvel and Rijnard (1949), who describe pseudotuberculosis in a baboon (*Papio papio*).

gut. Yet another case of *P. aeruginosa* in a simian presented a very different picture: an angwantibo (*Arctocebus calabarensis*) was found *post-mortem* to be very thin, with areas of consolidation throughout both lungs, whereas all other organs appeared normal. Bacteriological culture showed an almost pure growth of *P. aeruginosa*, and this had apparently caused pneumonia.

Owing to their dirty habits, all simians are potentially dangerous carriers of bowel-living organisms, and this will be considered in relation to *Salmonellae* and *Shigellae*.

To end the present section, attention must be drawn to the disease of melioidosis. This infection, caused by *Pfeiferella whitmori* (*Pseudomonas pseudomallei*), is an invariably fatal infection in man, which was formerly supposed to be confined to areas of the Far East, and to occur naturally only in wild rodents. It is usually transmitted by bites from infected animals. Recently a fatal case has been reported by Tammemagi and Johnston (1963) in an orang-utan in North Queensland, Australia. A further report has come to hand of a case of melioidosis in a monkey (*Macaca* sp.), kept in a zoo in Malaya (Retnasabapathy and Joseph, 1966).

The possibility that apes or monkeys imported from the Far East could carry this very dangerous organism is one that must cause a great deal of alarm. It is not even known, at present, whether infection in monkeys or apes causes active disease and is always fatal, or whether, as is the case with rodents, simian primates can be unaffected carriers. The latter prospect is uninviting.

Enteric infections of simian primates

Salmonellosis

No case of typhoid has ever been reported in any simian primate. The serum of one monkey, which died of enteritis at the London Zoo, agglutinated *Salmonella typhosum* antigen at a titre of 1:125. This was possibly a non-specific reaction and does not necessarily suggest that monkeys can be typhoid carriers. *Salmonella* infections indeed are not very common in primates other than man. An outbreak of food poisoning due to *S. aertrycke* was, however, reported in baboons at the London Zoo and was attended by a heavy death rate. The commonest *Salmonella* found in monkeys appears to be *S. typhimurium*, an organism commonly carried by mice, rats and birds. Man is susceptible to this *Salmonella*, and as with other diseases shared by man and monkeys, the connecting link again appears to lie with rodent populations in the vicinity of captive primates. *S. typhimurium* is more commonly found in South American species than in Old-World monkeys. There is little evidence that any group of lower primates can act as symptomless carriers of any *Salmonellae* for any length of time. It is far more likely that outbreaks of salmonellosis in apes and monkeys are derived from human sources rather than the other way round. It must, however, always be remembered that, unlike the shigelloses, *Salmonella* can exist in contaminated soil or vegetation, and infection between one species and another can therefore be indirect.

The views expressed above are derived from experience in the London zoological collections and are similar to those of Lapin and Yakovleva. These authors, describing experiences at Sukhumi, report two outbreaks of paratyphoid amongst monkeys during which a number of deaths occurred. The associated organisms were *S. breslau*, *S. paratyphosa* and *S. schotmülleri*. They quoted a paper by Aksenova and Khutsishvili (1949), describing a survey made on sections of the monkey house for paratyphoid carriers. During this survey, a small number of such carriers were identified. It is further stated that from 1957 onwards, when there were large shipments of monkeys from India and China, paratyphoid bacteria were again isolated. However, no overt cases of the disease developed, despite the fairly large number of bacillus carriers. In 1957, six monkeys were found to be carrying *S. paratyphosa*, and sixteen in 1958. At the end of 1958, eleven of three hundred rhesus monkeys were carrying *S. breslau*, *S. gaertner* and *S. paratyphosa*. The authors believed that the origin of the carrier state was contact with man at the place where the monkeys were assembled for transport to Sukhumi. This explanation would appear to be the most probable.

Aksenova (1958) later reported that the carriage of *Salmonellae* amongst Sukhumi monkeys was a constant feature during the period 1937 to 1945, reaching sometimes 16% to 21%. Different types were found in rhesus monkeys and baboons, but these were chiefly *S. typhimurium, S. paratyphoid B* and *S. paratyphosa* A.

Schneider *et al.* (1960), in a very extensive survey on the carriage of *Salmonella* and *Shigella* by newly received monkeys at the Okatie farms during 1955, found a different rate of incidence at different seasons of the year. Schneider's table analysing this work is reproduced with his permission.*

ENTERIC BACTERIOLOGICAL FINDINGS ON MONKEYS NEWLY RECEIVED AT OKATIE FARMS DURING 1955

Season	Rhesus*			Cynomolgus†		
	Animals exam.	Positive		Animals exam.	Positive	
		Shigella	*Salmonella*		*Shigella*	*Salmonella*
Winter	572	88 (15·4)‡	9 (1·6)	276	18 (6·5)	36 (13·0)
Spring	1143	146 (12·8)	52 (4·5)	293	13 (4·4)	47 (16·0)
Summer	1042	395 (37·9)	216 (20·7)	450	6 (1·3)	92 (20·6)
Fall	436	58 (13·3)	62 (18·1)	200	15 (7·5)	13 (6·5)
Total	3193	687 (21·5)	356 (11·1)	1219	52 (4·3)	189 (15·5)

* A total of 36,248 rhesus monkeys (8·8% sample size).
† A total of 24,261 cynomolgus monkeys (5·0% sample size).
‡ Per cent expressed in bracketed figures.

It will be seen that at certain seasons of the year the carrier state of *Salmonella* may reach more than 20%, and of *Shigella* more than 30%. The authors state that whereas *Shigella* infections tend to spread rapidly, to reach a peak and thereafter to decline slowly, *Salmonella* infections do not spread rapidly and generally disappear spontaneously. These remarks support the findings of Lapin and Yakovleva that salmonellosis is usually contracted, en route, from human sources, and vindicates the remarks made in the first paragraph to the effect that simian primates are not normally carriers of *Salmonella* organisms. Schneider *et al.* give the identification of their enteric organisms in a further table, also reproduced here.

* Dr Schneider's work was done on a grant from the National Foundation for Infantile Paralysis and from the Department of Defense, acting for the Commission on Enteric Injections, Armed Forces Epidemiological Board, Washington, D.C.

DISTRIBUTION OF ENTERIC PATHOGENS ISOLATED FROM 3250 MONKEYS (*) CULTURED ON ARRIVAL AT OKATIE FARMS

Genus	Type	No.
Shigella	*dysenteriae* 2	86
	flexneri 2	558
	flexneri 4	97
	flexneri 6	32
	sonnei	38
Total *Shigella*		811
Salmonella	*anatum*	126
	stanley	43
	typhimurium	42
	newport	23
	potsdam	9
	montevideo	4
	paratyphi B (var. *Java*)	3
	(other)†	6
Total *Salmonella*		256
Grand total		1067

* 2406 rhesus and 844 cynomolgus.
† *S. infantis, javiana, tennessee, darby, give,* and *oranienburg.*

Shigellosis

It appears then that *Salmonella* may infect primate colonies, and that the monkeys may either be seriously affected, or they may become temporary carriers. The most serious human pathogen in this group, *S. typhosa*, does not infect them, and in any case carriage of these organisms is in time self-limiting. It may furthermore be noted that there have been no reports of human infection or fatalities arising from simian sources. An important point, which should be noted, in relation to the epidemiology of the salmonelloses is that most species of salmonella can not only survive but multiply for considerable periods in moist vegetable material. Infection can be spread in this way, and it usually happens by the contamination of soil and vegetation with rodent or bird droppings.

Among the shigelloses, important differences must be noted, one being that no member of the dysentery sub-group of enteric organisms can survive for any length of time away from a vertebrate host.

The organisms most commonly associated with simian dysentery are the various members of the *Sh. flexneri* group, *Sh. sonnei* and *Sh. schmitzi*. A list of the serotypes infecting simians at the London Zoo is shown in the accompanying table.

SIMIAN PRIMATE DEATHS AT LONDON ZOOLOGICAL GARDENS ASSOCIATED WITH INTESTINAL BACTERIAL PATHOGENS

Simian species	Species of enteric pathogen
APES:	
chimpanzees	*Shigella flexneri*
gibbons	*Sh. flexneri, Sh. sonnei, Sh. schmitzi*
Langurs	*Sh. flexneri*
Guenons	*Sh. flexneri, Sh. sonnei*
Patas Macaques Baboons and Mandrills	*Sh. flexneri*
New-World Monkeys	*Salmonella typhimurium* *Shigella flexneri*
Marmosets and Tamarins	*Sh. flexneri* *Sh. sonnei*
Prosimians	
Galagos	*Sh. schmitzi*

Mrs. Anna C. Garner (1966, personal communication) has reported the isolation of *Sh. dysenteriae* from chimpanzees at the Delta Regional Primate Research Center.

From December 1965 to May 1966 *Sh. dysenteriae*, Covington, New Orleans isolated from the faeces of thirteen chimpanzees, of which ten have died of acute enteritis. There are no other reports of *Sh. dysenteriae* in Simians.

Shigellosis in simians often takes epidemic form and may pass quickly through the monkey house, causing serious losses. The dirty habits of most monkey species contribute to its spread; the animals defecate on the floor of the cage, drop their food on to the faeces, and then another member of the cage or a monkey from a neighbouring cage will retrieve the food and eat it. In this way, the disease can spread rapidly with serious consequences.

Apes and monkeys too are frequent carriers of organisms belonging to the dysentery sub-group; in most cases the organism isolated is *Sh. flexneri* (type 3), which is especially prevalent amongst gibbons. Chimpanzees and gorillas may also be affected and indeed any group of lower primates. Simians are also commonly carriers of *Sh. sonnei*, which may also cause diarrhoea and dysentery.

Monkeys infected with these organisms are a potential source of danger to man and there are records of serious tragedies which have happened as a result of contact with infected simians. Rewell (1949) described an outbreak of dysentery at the London Zoological Gardens amongst apes and their keepers, caused by *Sh. schmitzi*. The most serious outbreak of dysentery in humans arising from a simian source was that described by Bach *et al.* (1931), as a result of which seventeen cases of paradysentery occurred in humans following contact with a single shipment of pet monkeys, and three victims, all children less than five years old, died. This case is of such importance that Ruch reported verbatim an account of what happened. The same course is followed here:

> Five guenons, three adults and two young, were brought to a small German town from West Africa. Three of the monkeys had been sick, either in Africa or on the journey, with a dysentery-like disease, which had been treated with bismuth subgallicum. Shortly after the monkeys arrived, a fourth animal, an adult female caring for a young male, passed bloody stools.

This young male was given to the M family, and a day later he, too, passed bloody stools. A two-and-a-half-year-old child in this family then developed dysentery after eating a small cake discarded by the monkey. The child died after being ill two days, and shortly thereafter the remainder of the family, including four other children four to eight years old, became ill. Bacteriological examination showed *Sh. flexneri* type Y in the stools of four of the children. The female monkey died, and the young male was shot. Type Y bacilli were also found in the faeces of the female monkey and in sections of the intestines from both animals.

The surviving monkeys were isolated, but repeated examination of their stools did not reveal shigellae. Since these animals appeared healthy, children were again allowed to play with them. In the ensuing month ten more chidren, one-and-one-half to thirteen years old, fell ill of dysentery. Two, two and four years old, died. Stool samples from the surviving victims also contained type Y bacilli. The three monkeys were again isolated, although no illness was apparent. The owners later moved the monkeys to Hamburg. No subsequent outbreak of dysentery was known to have occurred there.

The children were the most seriously affected. The youngest child of the M family was sick only a few days before death, but in the other two fatalities, the deaths followed long illnesses. All the children suffered greatly. They were very hungry and thirsty, and passed stools (forty to fifty a day) which were chiefly bloody mucus, dark and foul-smelling. Also, tenesmus thrust forward the rectal mucosa, and there was a tormenting call to stool. The belly was distended and hard, and severe abdominal pain occurred after eating or drinking. The tongue was dry and cracked, and there was a persistent odour from the mouth. Complete apathy with deep somnolence developed; body temperature reached 40°–41°C. Recovery was slow, with many relapses. In contrast, the adult humans passed bloody stools for only a day, and there was no confinement.

Little could be learned about the illnesses of the monkeys. All had passed bloody stools at one time or another, and all had lost much weight before they arrived in Germany. Apart from the female which died (and the executed animal), the monkeys recovered fully. Necropsy of the two dead monkeys disclosed severe involvement of the intestines. The intestines and even the stomach contained shigellae. Major portions of the small intestines in these monkeys and in the fatally infected children were dark coloured.

The strains of bacilli were studied in culture. Serologic relationship between the strains and sera obtained from the sick people and the sick monkeys was established.

Although there was no bacteriologic evidence, it seems apparent that the surviving monkeys were the source of the second epidemic, especially since the infected children had played with the animals. It is unlikely that the original owners of the monkeys were carriers of the disease, even though they had had dysentery earlier; no bacilli were cultured from their stools. There was no evidence that any of the human victims could have been carriers, or that dysentery was dormant in other residents of the town before the monkeys arrived. The coincidence in place and time between the arrival of the monkeys and the outbreak of human disease suggests, therefore, that the animals had brought the disease with them. Absence of new cases during the period when the monkeys were isolated for study and after their departure is also suggestive.

How slight a contact between a monkey and a child may be to transfer disease is shown by the case of Carpenter and Sandiford (1952):

> One monkey grasped the child's ice cream cone, briefly touching the ice cream which the child then licked. Two days later the child was ill, and refused lunch. Diarrhoea began by evening, and two days later was mucoid and bloody. Two 6-day courses of succinyl-sulphathiazole were required to render the child's stools free from shigellae. The bacilli cultured from the stools were *Sh. flexneri* 103Z (=4b). As mentioned above, this type has frequently been the cause of bacillary dysentery in freshly imported simians, but has only once been reported from man in Britain. All the monkeys had been in the pet shop for nearly two years, and had been healthy and free from diarrhoea the entire time. After considerable effort, *Sh. flexneri* 103Z (=4b) was isolated from one monkey.

Carpenter and Sandiford stressed that if the organism had been one of the types common in man, the source of the infection could not have been traced.

More recently Robinson *et al.* (1965) described an outbreak of bacillary dysentery in Wales, which was traced to the importation of a pet monkey (probably *Cercopithecus mona*) from the Cape Verde Islands by a Merchant Navy steward. During the two months' voyage, the steward, who served food to the crew, and several members of the crew had diarrhoea, but at no time did the monkey show signs of illness. Shortly after the man's return to his home, of the seventeen persons at risk, seven children and one adult developed dysentery. *Sh. flexneri* (Y) was isolated from four children and the monkey. From the other three children *Sh. flexneri* (4a) was isolated, and the adult, who remained symptomless, was found to be excreting *Sh. flexneri* (Y) and *Sh. flexneri* (4a) on separate occasions. All the strains of *Sh. flexneri*, irrespective of serotype, had the rare and epidemiologically useful marker of being catalase-negative. This fact and the known antigenic relation of *Sh. flexneri* (Y) to *Sh. flexneri* (4a), combined with the sequence of the infections following the sojourn of the monkey within the different families suggested this to be a single outbreak, of which the Mona monkey was the carrier.

The chief carriers of *Shigellae* appear to be gibbons, and it is often stated that the bacteria are normal intestinal inhabitants. This is certainly not the view of the present writer. Infection often lurks and is symptomless, but at any time it may flare and cause serious symptoms, haemorrhage from the bowel and rapid death. The writer has found that the infection in gibbons, and indeed in other monkeys, is often located in the small intestine, and that it is then very difficult to isolate the organism in faecal cultures.

Carpenter and Cook (1965) examined rectal swabs taken from 370 Vervet monkeys and one hundred baboons immediately after their capture in Tanganyika and Kenya, but failed to isolate *Shigellae* from any animal. This is strong evidence that in this area at any rate shigellosis is not a natural disease in monkeys. However, the situation may be very different in parts of Asia, particularly where there are human–simian contacts. Russian writers report a high incidence of dysentery carriers amongst newly imported monkeys; for instance Abramova (1949) described spontaneous dysentery outbreaks amongst monkeys due to flexner bacilli. The infected animals were rhesus monkeys and baboons. They state that the incidence is highest in newly arrived monkeys in poor condition after transport, and that the disease occurs especially in severe winters when the monkeys' condition is poor. These remarks would indicate the probability of active disease developing from the carrier state. The same author (1952) reported a carrier rate of 9·8% amongst macaques and 6·2% amongst hamadryas baboons, the organisms involved being flexner bacilli of Z and Y type. Aksenova (1958) states that this dysentery was first isolated at Sukhumi in 1953, when more than one hundred strains were recovered from eighty-eight monkeys (rhesus, baboons and green). Prior to 1951, only dysentery flexner of various serological types had been isolated at Sukhumi, but during 1951–52, carriers of Schmitz bacillus and Sonne were also discovered.

Dzhikidze (1954) studied experimental bacillary dysentery in monkeys with special reference to the carrier state. She found that this could be created in rhesus, hamadryas baboons and green monkeys. Dysentery carriers spread amongst all groups, particularly the rhesus monkeys, which are most susceptible. She considers that symptomless infected monkeys are the reservoir from which spread occurs, but that monkeys which regularly carry dysentery bacilli show some clinical and morphological signs. These arguments are developed by the same author (1956), when she shows that the so-called healthy carrier animals (rhesus monkeys and hamdryas baboons) in fact show the signs of low-grade infection, and bowel lesions can be demonstrated by endoscopy. The Russian results are further reviewed by Lapin and Yakovleva, who quote a further interesting series of data in this connection. During 1957, four hundred Indian rhesus monkeys were delivered to the animal house at Sukhumi between May and December. 76·5% of the animals (306–400) were clinically sick, as revealed by the presence of diarrhoea of different kinds and severity. Dysentery bacilli, chiefly of the flexner type, were found in the faeces of 256 animals (64%). Within the first two months of their arrival, 124 (31%) died. During the same period of time the main group of six hundred monkeys in the animal house included 195 carriers or 32·5%. Of these, twenty-six animals (or 4·3%) were sick, and thirteen eventually died of the disease.

Reference has already been made in the foregoing section on salmonellosis to the paper by Schneider *et al.*, and their tables of infection rates with *Salmonellae* and *Shigellae* have been reproduced above. It is evident that bacillary dysentery, arising from all groups of *Shigellae* is of very great importance in simian colonies, both in relation to the health of the animals themselves, and as zoonoses from which serious diseases may arise either amongst keepers or amongst children who keep the animals as pets. The situation with regard to the shigelloses is potentially so serious that it must reinforce the arguments for proper screening by laboratory tests of all primates before they can be accepted as pets and handled by children. Failure to adopt these precautions may result in child deaths from bacillary dysentery and persons not taking such care are guilty of gross negligence. The recent discovery of *Sh. dysenteriae* in chimpanzees adds to the force of these remarks.

Proteus

Proteus morgani, believed to be the cause of summer diarrhoea in children, also causes disease, sometimes a severe type, in simian primates. Lovell (1929, 1930) at the London Zoo, isolated Morgan's bacillus from the faeces of rhesus monkeys suffering from varying degrees of enteritis. Enteritis caused by this bacillus has also been described in London by Hamerton (1929, 1934). Pathogenesis of the disease in monkeys has been reviewed most extensively by Lapin and Yakovleva (1963), who state that this is a common cause of death in young monkeys. The course of the disease as described by these workers is marked by sharp deterioration in condition, sluggishness and rejection of food, flatulence, vomiting and diarrhoea. Morgan's bacillus was recovered from the faeces and proved to be pathogenic in white mice. There are insufficient data to state whether a carrier state of Morgan's bacillus can exist. It is obvious, therefore, that whether monkeys are carriers or not, they could transmit this infection to children who were in contact with them.

Tuberculosis

Tuberculosis in simian primates may be due to human, bovine or avian strains of mycobacterium, but there is almost no recorded information as to the type of disease caused by the avian bacillus. The bovine bacillus usually causes a chronic type of disease, in which the mesenteric glands and other abdominal organs are involved. Tuberculosis of the bowel itself is, however, rare in simian primates. The bovine type of disease often appears to be regressive, and natural recovery may take place. Amongst primate collections it may show little effect, and become active when conditions are adverse or the weather is cold. For this reason it is difficult to eradicate, especially since diagnostic methods are somewhat unreliable.

Human-type tuberculosis in simian primates is classically described as a fulminating disease of the respiratory type with a course of at most a few months and ending invariably in death from generalised tuberculosis. This assessment of the disease appears in general to be correct; and this being the case, monkey infection must in most cases arise from human sources, and the monkeys are a lesser danger to man except during the short course of the infection. Even so, infection can hardly exist unobserved in a monkey colony, and steps should be taken quickly to control or eliminate it.

The classical description of tuberculosis due to human strains of mycobacterium in the lower primates is however not universally acceptable, and a study of the records in London shows that it is only true in certain groups. The great apes, the African and Asiatic monkeys are very susceptible and the course of the disease in them is usually as described. Tuberculosis of any kind is extremely rare in all New-World species, and it is not commonly encountered amongst pro-simians, such as galagos and lemurs. When tuberculosis occurs in these groups, it resembles more the bovine type of disease in the more susceptible genera. It is usually of the abdominal type, and lesions in the mesenteric glands may appear to be regressive. These animals, though unlikely to be infective at this stage, could harbour the disease undetected and might in the end be of greater danger to human contacts than those animals which showed more active symptoms likely to result in rapid death.

BCG vaccine is widely used to protect apes and monkeys against tuberculosis and it appears to be helpful. It has the additional advantage that, when used on an infected animal, it provokes a violent reaction and is in this way probably the most reliable method of tuberculin testing. Tests with standard tuberculins, though often very helpful, are unreliable, and a negative result does not preclude the possibility that the animal may be infected. Chest X-rays may also be used, but, where positive, the monkey is likely to be so severely affected that it will not survive long in any case. Examination of the faeces for the presence of acid-fast bacilli has been used in London as an adjunct to diagnosis for a number of years. Comparative tests have shown this to be more reliable than the examination of sputum. As much faeces as can be collected are divided into two portions, one of which is digested at incubator temperature in antiformin, the other in acid phosphate. After twenty four hours at 37° C both portions are then centrifuged. The sediment from the antiformin digest is stained on a slide with Ziehl-Neelsen and examined for acid-fast bacilli; the sediment from the acid phosphate digest is neutralised and used to make cultures on modified Dubos and Löwenstein's media. The Dubos culture may be positive in six to eight days and the Löwenstein in two weeks.

Care needs to be taken in the interpretation of these results, since acid-fast bacilli, which are not *Mycobacterium tuberculosis*, are sometimes found in simian faeces. These should be transferred to ordinary culture media, and attempts made to decolourise them with both alcohol and alkali as well as acid. By these means a differentiation may sometimes be made, but even so harmless mycobacteria may still remain. It is therefore necessary to assess the significance of a positive faecal sample in the light of other tests and of clinical findings.

The treatment of simian primates for tuberculosis appears to be less successful than that of humans and relapses have been reported. An ape or monkey infected with tuberculosis is a dangerous animal and, except in exceptional circumstances, should be destroyed. It should never be retained as a family pet, though cases may occur where retention in zoological collections under proper precautions might be justified.

Reference has already been made to the fundamental work of the late Sir Harold Scott (1930) in connection with mixed mycotic and tuberculous infections. Scott, who was for some years Pathologist to the Zoological Society of London, later became Director of the London School of Hygiene and Tropical Medicine. In the latter capacity he correlated his extensive notes on the occurrence of tuberculosis in both man and simian primates and produced what is still the classical description of this disease in apes and monkeys. His conclusions that tuberculosis in simian primates is analogous to the disease as it develops in children is highly relevant to our enquiries into these diseases as zoonoses. In this connection, Scott's work, undertaken in collaboration with Professor John Beattie, was guided by a search for the primary site of the disease, the portal of entry of the bacillus, and for a rational explanation of the mode of spread. He concluded that the respiratory tract was the commonest portal of entry and the lymphatic system the chief route of extension. He therefore dissected the lymphatic system in different groups of monkeys, making the original observation that in the New-World monkeys there is no true thoracic duct: the lymph from the abdomen and hind end of the body is collected into trunks which open into the venous system in the region of the junction of the renal veins with the inferior vena cava.

In this work he was struck by the remarkable similarity which the distribution of lesions in simian primates bore to the same disease in children. The ready susceptibility of children in early life bears a close analogy to that of an unimmunised race of animals, when removed from their natural environment to circumstances in which they might be exposed to infection from human cases of the disease or from an infected member of their own group, accidentally imported with them. Scott's observations on the distribution of the lymphatic vessels were based on dissections of gorilla (1), chimpanzee (3), orang-utan (1), gibbon

(2), *Cercopithecus* spp. (10), *Macaca mulatta* (3), *M. irus* (3), *M. nemestrinus* (3), *M. silenus* (1), *Presbytis entellus* (1), *Colobus polykomos* (1).

Scott gives a detailed analysis of his conclusions and compares *post-mortem* notes on simian cases with corresponding cases amongst human children.

Fox noted – as has the present author – the varied susceptibility of different simian groups. He found that amongst the Old-World monkeys the numbers attacked were two and a half times as great as in New-World groups. Amongst thirty-two marmosets examined by him no cases of tuberculosis were found.

The importance of tuberculosis as a human disease amongst advanced communities has diminished so spectacularly that some may be inclined to dismiss it as of little significance. Such a view is erroneous indeed, because many imported simians, particularly the macaques from India and other parts of the Far East, are derived from areas where not only is the incidence of tuberculosis still high, but simian–human contacts occur much more frequently than in other parts of the world. Furthermore, where precautions over Mantoux testing and BCG immunising have been neglected, children may well be much more susceptible to the disease than would have been the case in previous generations. There thus exist grave dangers of importing infected simians, and an infected animal could spread disease to susceptible children or even adults with very serious consequences. As with *Shigellae*, it is criminal folly to admit newly imported monkeys to the companionship of children until they have been thoroughly tested, and it is virtually certain that they are free from tubercle infection.

Little has been said of possible danger from simian primates, which carry bovine or avian tuberculosis. The latter may be dismissed as unimportant. There is little in the literature on the subject of bovine tuberculosis in simian primates. This, in the experience of the author, is a relatively benign form of the disease and is often found in apparently healthy monkeys of chiefly Old-World species, which possess indurated or caseous mesenteric glands. Under poor conditions of cold or ill-health from other causes, such glands occasionally break down, and generalised tuberculosis may ensue with fatal consequences. Most monkeys and apes die so quickly from human tuberculosis that the period during which they cause risk to humans is limited. Possibly the less severe bovine tuberculosis, during the course of which acid-fast bacilli are occasionally excreted in the stools, may be a more serious hazard.

At the present time (February 1966) there has been no human tuberculosis amongst the primates at London Zoo for some five to six years. It was eventually eliminated, but sporadic deaths from tuberculosis of the bovine type have continued until a year ago. These have all occurred with the same picture of breakdown of indurated mesenteric glands, from which generalised fulminating disease has arisen. During this time, all the simian primates have been regularly tested, both by tuberculin tests and by faecal examinations. These tests have all been negative and there has therefore been no way of knowing which animals have been infected. Nevertheless, there is no evidence that any spread has occurred in the monkey house. For the tuberculin test, BCG injections are used, since these give positive reactions in animals infected with human-type bacilli and have the added advantage of giving some immunity. Obviously these tests are not fully satisfactory; in addition, to strict screening of newly imported monkeys which are to come in contact with children, care should be taken to see that the children themselves are properly protected with BCG vaccine, if found desirable on the results of the Mantoux test.

Miscellaneous Organisms

No mention has been made of *Listeria*, *Erysipelothrix*, mycoplasmas, rickettsias or organisms of the Lymphogranuloma-Trachoma group.

Heisch *et al.* (1962) found antibodies to *Rickettsia conori* and *R. burnetti* present in the sera of many species of rats caught in Kenya, and they also tested baboons, mongooses, an ichneumon genet, serval cats, a palm civet, a leopard and a hyaena. All these animals contained rickettsia antibodies, usually in low titre. There were more sera positive for *R. burnetti* than for *R. conori*. Three baboons, *Papio doguera*, gave titres of 1:100 for *R. burnetti* and the palm civet reacted at 1:160. Other workers have also found evidence to show that simian primates do become infected with Q fever in the wild, and they may, in common with almost all other wild animals, be regarded as possible carriers.

There is no evidence that any simian group normally suffers from trachoma in the wild; they can, however, be readily infected as shown by Nicolle and Lumbroso (1926) who inoculated human trachoma virus into a number of laboratory animals. The virus caused granular conjunctivitis in *Papio* sp., and subinoculation to another animal of the same species also caused this disease. There are many other such experiments proving that apes and monkeys are susceptible, but since trachoma can hardly be regarded as an important zoonosis, it is not worth while to review the literature.

There is at present no information about infection of simian species with the other organisms mentioned above, and no evidence that these could be primate zoonoses. It is surely to be expected, however, that apes and monkeys will carry, and be infected by, *Listeria* from time to time, and that they can carry mycoplasmas.

Part IV Viruses

9 Viruses

A General

It has until recent times been a matter of some difficulty for a non-virologist to discuss diseases of virus origin with any degree of intelligibility, because of the lack of any agreed system to classify these organisms. This difficulty has been overcome with the publication of Sir Christopher Andrewes' (1946) *Viruses of Vertebrates*, from which the classification given in the table is taken. A list of those virus groups with the subheadings which are of importance to our subject is also included.

For these purposes, the subject of viruses can be classified under four clearly defined sub-headings:

1 Virus diseases of the arboreal disease cycle, which may give rise to epidemics if the arboreal and terrestrial disease cycles become entangled. Such diseases are solely due to arbo-viruses, and are typified by yellow fever, dengue and Kyasanur Forest disease.

2 Those diseases, from which simians may suffer only mild symptoms, but which may be much more dangerous in humans. Diseases of this type are typified by B-virus (*Herpesvirus simiae*), infectious hepatitis and others.

3 Human diseases, which may be passed from man to monkeys (anthroponoses) and which monkeys may retransmit to man, thus acting as an unwitting intermediary in the spread of what is really a human disease. Simians may play a part of this kind in the spread of many common human diseases, such as measles. Both apes and monkeys have been extensively used in research on human diseases, and it may be supposed that they could play this part in relation to any human disease in which it can be shown that they become infected by natural routes of transmission. One cannot regard as zoonoses dis-

eases which can be transmitted to simian primates only by unnatural routes such as by intracranial infection or infection of unborn foetuses.

4 Those viruses which naturally occur in apes and monkeys, and are apparently non-pathogenic, but might conceivably cause disease in human beings by being transmitted in immunising agents manufactured from simian tissues. The outstanding example of this is the SV-40 virus, which must have been administered to thousands of humans in polio-virus vaccine – fortunately without undesirable effects – but which causes sarcoma in hamsters. The term SV-40 refers to simian virus 40 of the simian-virus series introduced by Hull and his collaborators (1956, 1957, 1958). This series is simply a list of the many simian viruses recovered from monkey kidneys or stools in the course of research work on poliomyelitis and other diseases, most of them being 'orphan', which means that they are simply viruses which seem to have no pathogenic effects.

For the sake of clarity, the Arbo-viruses will be considered first, outside their natural place in the scheme of classification, because this is their natural position in relation to the scheme given above. Thereafter, the important viruses will be considered in the order and the scheme of classification used by Andrewes. Whereas with bacteria and other organisms, one can be reasonably certain that an adequate coverage of the subject has been given, this is by no means the case with viruses. Those viruses which, at present, appear to be important will be given reasonably full treatment; others considered to be of potential importance will be mentioned rather more briefly. Undoubtedly many of the latter and many of those not mentioned – perhaps unknown at this time – will eventually prove to be of great importance. Those people concerned with the potential danger of transmitting virus diseases to man must be continually aware of the dangers of some new thing. '*Ex Africa, semper aliquid de novo*!'

B Arbo-viruses

More than one hundred and fifty arbo-viruses have been isolated, and more are added to the list yearly. Most of them are infectious to man, the majority are mosquito-transmitted, and in a great many cases monkeys act as the primary hosts. The table given of the most important groups is taken from the *International Symposium* (1962) *on Comparative Medicine*.

Arthropod-borne viruses are of the greatest importance in tropical countries where conditions are especially suitable for the vectors. The habitats of simian primate groups are mostly located in such areas, and it is not surprising therefore that they are naturally infected with many such agents. The arbo-virus which has been most studied is that which causes yellow fever, the life history of which is therefore well known. Since this disease provides a satisfactory model of the ways in which such viruses can be spread and come to infect human populations, and of the dangers associated with them, a brief account of the bionomics of yellow fever will be given.

Yellow fever is properly regarded as being caused by a commensal virus found in mosquitoes inhabiting the upper canopy of the evergreen forests of Central tropical Africa. In Africa the virus is specific to mosquitoes of the genus *Aëdes*, and the common carrier is *A. africanus*. Monkeys inhabiting the upper forest arcades become infected by the mosquitoes at an early age; they pass through a mild disease then develop immunity, following which the virus is eliminated. The virus cannot be isolated from recovered monkeys, although antibodies can be shown to be present in the blood serum. It is, therefore, the mosquito which forms the reservoir of virus and not the monkey. The human disease is transmitted by another mosquito, *A. simpsoni*, which acquires the virus from monkeys raiding banana plantations. This, being a ground-feeding mosquito, attacks man and is thus able to introduce the disease to the human population, in which infection is maintained by the domestic mosquito *A. aegypti*. The disease in man is attended by a high death rate.

Yellow fever was possibly transported to the American continent by means of infected mosquitoes lurking in ships. In the forests of South and Central America, it found a habitat ideally suited to it and genera of mosquitoes (*Sabethes* and *Haemagogus*) in which it was able to form a further reservoir. Unlike the African

PROVISIONAL VIRUS CLASSIFICATION

Nucleic acid	RNA				DNA		
Symmetry	Cubical		Helical	Uncertain	Cubical		*Uncertain symmetry
Presence of outer membrane	0 (ether stable)	+ (ether labile)	+ (ether labile)	+ (ether labile)	0 (ether stable)	+	+
	Picorna-virus Reo-virus	Arbo-virus (symmetry not certainly established and group may be heterogeneous)	Myxo-virus Rabies	Fowl-leucosis complex Viruses associated with mouse tumours	Adeno-virus Papova-virus	Herpes virus (ether labile)	

PRINCIPAL ARTHROPOD-BORNE VIRUSES TRANSMISSIBLE TO MAN

Virus	Casals' Group	Distribution	Suspected Vectors	Probable Principal Hosts
Chikungunya	A	Africa	*Aëdes africanus, aegypti*	Monkeys, man
Eastern encephalitis	A	E. North America to Brazil; Dominican Republic	*Aëdes* spp., *Culiseta melanura*	Birds
Mayaro	A	Trinidad, Brazil	Mosquitoes	Unknown
Uruma	A	Bolivia	Mosquitoes	Unknown
Semliki Forest	A	E. Africa	*Aëdes* spp., *Eretmopodites* spp.	Monkeys, man
Sindbis	A	Egypt, W. Africa	*Culex univitatus, antennatus, pipiens*	Birds? Man?
Venezuelan encephalitis	A	Trinidad, S. America	*Mansonia* spp., *Aedes* spp.	Mammals
Western encephalitis	A	W. North America to Argentina (E. North America)	*C. tarsalis, Aedes* spp.? (*Culiseta melanura*)	Birds
Dengue	B	Trinidad, C. and S. America, S. Pacific, Orient	*A. aegypti, albopictus*	Monkeys, man
Iheus	B	Trinidad, Guatemala, Brazil	*Psorophora* spp.	Mammals
Jap B encephalitis	B	Orient	*Culex tritaeniorhynchus, pipiens, quinquefasciatus*	Birds
Kyasanur Forest	B	India	Unknown	Unknown
Murray Valley encephalitis	B	Australia	*C. annulirostris*	Birds
Russian spring–summer	B	Central Europe, Russia, Siberia, Orient	Ticks (*Ixodes* sp.)	Mammals (rodents?)
St Louis encephalitis	B	C. and W. United States, Haiti, Trinidad	*C. tarsalis, pipiens, quinquefasciatus; Culex* spp.	Birds
West Nile encephalitis	B	W. Africa, Egypt, Israel	*C. antennatus, univitatus, pipiens*	Birds, man
Yellow fever	B	C. and S. America, Trinidad, Africa	*Haemagogus* spp., *Sabethes chloropterus, A. africanus, simpsoni, aegypti*	Monkeys, man
Sandfly fever	Ungrouped	Mediterranean region	*Phlebotomus papatasi*	Man

monkeys, those of South America (Cebidae, *Aotus*, *Alouatta*) proved susceptible to the fever which occurs in epidemic form among them, sometimes causing high mortality. Epidemics in the monkey kingdom frequently coincide with epidemics in man. The transmitting agent in man is *Aëdes aegypti*, and it appears that infection passes from monkey to man when trees are felled in the forest, and numbers of infected mosquitoes are brought to ground level. Woodcutters thus become infected and pass the infection to *A. aegypti*, which transmits the disease from man to man in settlements and townships.

In view of the ease with which yellow fever appears to have crossed the Atlantic to the South American continent, it is odd that the disease has not spread eastwards into the monkey populations of Asia. It has been suggested that the Asian equivalent of yellow fever is another mosquito-transmitted arbo-virus disease, namely dengue, and that yellow fever and dengue cannot co-exist because of cross-immunities. Thus, dengue, common throughout the Far East, may be regarded as an Asian mutant of the yellow-fever virus, and in these areas may take its place as the epidemic arbo-virus zoonosis of primates. Chimpanzees and most species of monkeys are only inapparently affected, except by mouse-adapted strains, Mosquitoes of the *Aëdes* group, as with yellow fever, form the main reservoir of the virus, which again as with yellow fever, is transmitted from mosquito to mosquito from infected monkeys or other jungle animals.

During the work on yellow fever at the Yellow Fever Institute of the Rockefeller Institute at Entebbe, Uganda, a number of previously unknown viruses were isolated, some subsequently found to be of considerable importance. Such viruses were West Nile, Bwamba fever, Semliki Forest, Ntaya, Bunyamwera, Uganda-S and Zika. Since this Institute was taken over by the East African Virus Research Institute, more viruses have been isolated and investigated. The work of both institutes has been reported in numerous papers in the *American Journal of Tropical Medicine*, the *Journal of Infectious Diseases*, the *Annual Reports of the East African Virus Research Institute* and in other journals. It will serve little purpose to give a full account of all those viruses which can cause disease in man and monkeys, indeed the pathogenicity of many of them in simian primate hosts and their incidence in wild monkeys is unknown. Possible dangers arising from the sudden appearance of new viruses by mutation, or by changed transmission cycles from primates living virtually isolated in the forest canopy, can be well illustrated by reference to two new human diseases which have become widespread in recent years, one in Africa and one in India. These are O'nyong-nyong fever and Kyasanur Forest disease.

Early in 1959 a new disease, O'nyong-nyong fever, appeared in human populations in North-west Uganda and spread across the northern half of the country in a south-easterly direction. Infection was accompanied by headache, fever, swollen glands and crippling joint pains; it was followed on the fourth day by the eruption of an itching rash mainly on the body and arms. These symptoms persisted for about five to seven days and were generally followed by quick and complete recovery. During the next two years the disease spread with great rapidity throughout Uganda, Kenya, Tanganyika, Nyasaland and even reached the Rhodesian borders. Fortunately the disease remained mild, but it illustrates forcibly what could happen if a new virus disease arose and caused a heavy death rate.

O'nyong-nyong is a virus disease transmitted by two species of mosquitoes, *Anopheles gambiae* and *A. funestus*, and possibly also by bed bugs. The virus is supposed to be a new one, which arose by mutation. If so, it may have arisen from a virus which was first isolated from human patients during the period 1952 to 1953 and named Chikungunya (Chik). This disease occurred in Tanganyika and in 1958–9 in the North-eastern Congo. The same virus was isolated in 1956 from specimens of the mosquito *Aëdes africanus* taken in Uganda sixty-five feet above ground level in lakeshore forest. The relationship between O'nyong-nyong and Chik is unproven, as also is its possible origin from monkeys; nevertheless this possibility is alarming, and the consequences could have been serious if the virus had caused severe illness in the human population. It shows how, at any time, a highly lethal mutant virus could arise and could cause the death of a large proportion of the human population.

Kyasanur Forest disease was first discovered in late 1955, when a large number of bonnet macaques (*Macaca radiata*) and langurs (*Presbytis entellus*) were found dead in Kyasanur Forest in South-western India. At the same time, there occurred a severe prolonged febrile illness, called 'the monkey-disease' by the natives, in a number of humans living in the area. The outbreak lasted three months, and a second outbreak began early in 1957 over a much wider area. Investigations showed that the disease in both monkeys and man was caused by a virus (KFD) of the Russian spring-summer virus complex. This is transmitted by ticks of the genus *Haemaphysalis*, of which many species occur in the area as parasites of both man and monkey. Virus was isolated from ticks by Trapido *et al.* (1959). *Haemaphysalis* are three-host ticks; at each moult from larva to nymph and from nymph to adult the tick falls to the ground and acquires a new host. Larval and nymphal stages are parasitic on arboreal species, such as monkeys, the adults attack ground-living ungulates. The disease in monkeys thus shows a seasonal incidence, being present in the dry weather when the ticks are in the larval and nymphal stages. During the wet season very few ticks are found on the monkeys, although cattle and wild ungulates become heavily infested with adult males and females.

Kyasanur Forest disease illustrates the possible dangers of virus life cycles which may arise by means of ticks which are dependent upon different hosts at different stages of their life cycle, particularly if they come to parasitise unusual hosts. For an account of the biology of ticks, see pp. 39–41.

These instances must be regarded as illustrative of potential dangers from epidemic diseases which arise from the existence in a contiguous but separated habitat of animals so closely related to man as are the apes and monkeys.

C Picorna-viruses

The term Picorna-virus was adopted by the *Report of International Study Groups* (1963) to cover the very small, ether-resistant

RNA viruses. It includes the entero-viruses, including poliomyelitis, Coxsackie and ECHO. It also covers miscellaneous viruses, such as foot-and-mouth disease, encephalo-myocarditis (EMC), some other animal viruses and also the rhino-viruses, which cause the common cold.

Polio-viruses The evidence with regard to natural infection of both monkeys and apes with all three strains of polio-virus is indecisive, though both are normally resistant to infection except by the cerebral route in experimental procedures. Andrewes states that chimpanzees and cynomolgus monkeys are readily infected *per os*, and chimpanzees, unlike lower primates, can readily be made into symptomless intestinal carriers. Ruch gives some circumstantial evidence showing that both macaques and chimpanzees have become naturally infected with polio viruses in use in laboratories and that chimpanzees, at any rate, have been infected in zoos by contact with children in the incubation stages of the disease. There appears to be no evidence that any infection of man has occurred as a result of contact with any simian primate. Poliomyelitis is therefore not an important primate zoonosis, although it would be wise for children who play with monkeys or apes to be immunised against it.*

Coxsackie virus group There is some evidence that members of the A strains of this group show some neuro-pathogenicity for monkeys. The Coxsackie group contains at least thirty viruses, which cause diverse diseases, such as meningitis, epidemic muscular diseases of rheumatic type and diseases of the heart in the new-born. In human beings, these diseases chiefly affect children and may cause fatal illnesses in the very young. There is some evidence that this group of viruses may infect macaques, which may carry disease in silent form.

The ECHO viruses The ECHO viruses can all cause mild disease in man, such as diarrhoea, mild respiratory-tract infections and colds. Some strains also cause a form of meningitis, the symptoms of which may resemble poliomyelitis. They are sometimes found in faeces of apparently normal children, and monkeys

* *Vide addendum* (p. 152), which reports two cases of suspected naturally occurring poliomyelitis in gorillas at the Yerkes Regional Primate Center.

carry them in the bowel, where they are known to virologists as 'simian foamy agents'. All members of the entero-virus group multiply predominantly within the cells of the intestinal tract, and the ECHO sub-group is usually carried in silent form by monkeys.

Rhino-viruses Evidence of the susceptibility of monkeys and apes to common-cold viruses is rather tenuous, as it is also for the influenza viruses. It is, however, true that chimpanzees commonly share in epidemics of the common cold, and that the disease is mild and recovery takes place in a few days. They do, however, tend to contract secondary infections from which pneumonia, sometimes serious, may result. There seems little doubt that chimpanzees can acquire the infection from humans, and humans from chimpanzees. Monkeys, as opposed to apes, do not appear to be susceptible to common colds.

D Reo-viruses

Numbers 4, 12, 28 and 59 of the simian virus series appear to be reo-viruses. Sabin (1960) described an outbreak of coryza in chimpanzees, associated with steatorhoeic enteritis, as being due to a Type 2 reo-virus. Type 2 (SV-59) has also been isolated from chimpanzees and from a monkey which died of pneumonia. Type 1 (SV-12) has been isolated from both rhesus and cercopithecus monkey kidneys (Malherbe and Harwin, 1957). Not a great deal is known about this group of viruses in relation to human disease, but Types 1 and 3 have been isolated in cases of mild fevers with diarrhoea in children. Whether or not the viruses have caused the illness is uncertain. There is therefore reason to believe that chimpanzees or monkeys may be carriers of viruses of this group, and that they might transmit them to children. The diseases thus caused are mild. The infection, as in adult humans, is usually inapparent.

E Myxo-viruses

Influenza A, B, C It is generally supposed that apes and monkeys suffer from influenza acquired from human sources. It is undeniable that, during the course of human epidemics, apes in particular suffer from respiratory troubles which appear to be similar to those affecting the human population. There is, however, little evidence that these diseases are in fact caused by influenza viruses, and indeed what little evidence exists seems to point to the contrary view. The question has been well summarised by Ruch, from whom the following remarks are quoted:

> Old-World monkeys are normally quite resistant to influenza; the susceptibility of New-World species has not been investigated. Macaques inoculated intra-nasally with virus A (strain BR 8) by Woolpert *et al.* (1941) did not develop the clinical signs of the diseases, but reacted with an antibody response and changes of the blood picture. Clinical signs were also minimal when the influenza virus and *Streptococcus haemolyticus* group C were given together or sequentially (Merino *et al.*, 1941).
>
> On the other hand three of four macaques kept at environmental temperature between 4° and 6°C died after inoculation with influenza virus (Saslaw *et al.*, 1946). The influence of low temperature was presumed to be a lowering of the monkey's normally high body temperature to the optimal range in which the virus is effective – i.e., to the temperature of man, who is susceptible.
>
> The recent isolations of previously unknown viruses throw some doubt on the diagnosis of 'influenza' for spontaneous illness when the cause is not identified virologically, immunologically or pathologically.

Ruch goes on to criticise the diagnosis of influenza in South American monkeys and chimpanzees where the virus was not positively identified. It is evident that virus identifications need to be made urgently in all cases of respiratory disease in monkeys.

Mumps When experimentally introduced, the mumps virus causes disease in rhesus monkeys and other primates. However, no certain identification in naturally occurring cases of mumps have been described in apes or monkeys. Osman Hill (1959), in Ruch (1959), p. 72, described a supposed case in a *Cercopithecus neglectus* which recovered. A further possible case was investigated by the author in a galago, which suffered symptoms of parotid-gland swelling after being in contact with an infected child; the galago died, but no virus was recovered from its tissues at the Public Health Laboratories at Colindale. Parotitis in monkeys, as noted under, may frequently be associated with cytomegalo-virus infection.

Measles Whether or not measles is an important disease in wild apes or monkeys is not known, and neither is there any indication of whether simians have any great importance in the epidemiology of the disease as a zoonosis. It seems improbable that their role in transmission could be of importance; however, it is clear from the work of Russian observers that monkeys (rhesus and hamadryas baboons) can be experimentally infected with measles and can develop immunity to the disease. Konova *et al.* (1945) infected fifty-six hamadryas and rhesus monkeys intravenously, by way of the upper air passages, and by both routes simultaneously. In 56% of the animals, clinical symptoms were those of typical measles with active immunity to reinfection. Konova and Ponomarova (1949) tested the duration of immunity in rhesus, hamadryas and hybrid macaques and found this to be not less than seven and a half years; the period of observation. Reazantsova (1956) reported a naturally occurring outbreak of measles amongst the monkeys of the Sukhumi animal house. The epidemic involved thirty-one monkeys of different species, which all showed different susceptibilities to infection. Rhesus monkeys and hamadryas baboons were the most susceptible.

Whereas, there is no doubt, monkeys and apes can be and sometimes are readily infected by humans with measles, it seems doubtful whether they could play an important part in the spread of the disease. It is, however, possible that, during the course of a measles epidemic, children might be infected by visiting the monkey house at the Zoo. During cases of known outbreaks of measles amongst monkeys, the animals should probably be placed in quarantine just as in-contact susceptible children are.

Syncytial (RS) virus Syncytial virus is also known as chimpanzee coryza agent. The virus causes outbreaks of colds amongst captive chimpanzees and it can also infect ferrets, mink and marmosets. The infection is prevalent over brief periods during the winter months. In man, this virus causes respiratory infections with rhinitis and coughing. In children, especially infants, it can cause bronchitis, bronchiolitis and pneumonia, and can sometimes prove fatal.

F Rabies

Monkeys can be infected experimentally with rabies, but there is no recorded incidence of rabies being transmitted to a human being by the bite of a monkey. Nevertheless, the possible danger that a newly imported monkey might be in the incubation period of rabies should be kept very much in mind.

South American species, which live in areas where vampire bats (family Desmodontidae) are also present are obviously dangerous if some of the bats are carrying infection. Accounts of conditions in South America in the upper Amazon Basin, such as those of Fawcett (1953) in his *Exploration Fawcett*, suggest that both simian species and vampire bats are present in very large numbers in arboreal habitats. Vampire bats are eternally seeking prey. It would appear from published accounts that infected vampires are mostly present in the savannah regions of South America, where they attack cattle and cause considerable losses amongst them as a result of the rabies they transmit. There do not appear, so far, to be accounts of infected vampires in forested regions, which is probably why no monkeys have so far been described as suffering from rabies. The disease might obviously be spread to forest-living vampires, and a serious situation could develop for the simian populations. This could be aggravated if the fruit-and-insect-eating bats in the area were also to become infected, as in other parts of Central and South America and the USA, particularly since instances are now known of rabies transmission by aerosol means. The rabies situation is evidently one that requires to be carefully watched, and for this, as for other reasons, newly acquired monkeys should be kept under strict observation in quarantine before being too freely handled or admitted to the companionship of human beings as pets.*

G Adeno-viruses

Some seventeen adeno-viruses are included in the SV series and share complement-fixation antigens with human strains but they

* *Vide addendum*, in which a naturally occurring case of rabies in an imported *Mocaca mulatta* is described.

are shown to be distinct by neutralisation tests. Only one of these (M.6) has been found pathogenic, causing rhinitis and conjunctivitis in captive patas monkeys (*Erythrocebus patas*).

H Herpes-virus group

There are four described species of herpes virus:

1 *Herpes-virus simplex;*
2 *Herpes-virus simiae* (B-virus);
3 marmoset virus (*H. tamarinus* or *H. platyrrhinae*);
4 *H. zoster* (varicella).

Herpes simplex is a benign skin disease of man and is of no significance as a zoonosis.

Varicella (chickenpox) has been described by Heuschele (1960) in chimpanzees, orang-utan, and gorilla. The disease is mild and of little significance as a zoonosis.

The *marmoset virus* is a herpes-like virus isolated from throat swabs and autopsy material of marmosets (*Callithrix* and *Leontocebus* spp.) (Holmes *et al.*, 1963; Deinhardt and Deinhardt 1966). The virus causes an invariably fatal disease in marmosets, although it appears to be a natural pathogen of spider and squirrel monkeys (*Ateles* and *Saimiri* spp.). For this reason, Deinhardt is proposing a more appropriate name of *Herpes-virus platyrrhinae*. In view of the latest report on the isolation of this virus from a number of Douroucoulis, or owl monkeys (*Aotus trivirgatus*) in which it caused death in a large percentage of newly imported animals (Hunt and Melendez, 1966) this name would certainly appear to be more suitable, as the virus does not seem to be confined to marmosets only. There have been no cases of disease occurring amongst persons who have handled any of the infected species, and it is believed to be non-pathogenic for man. Immunity to *H. simplex* appears to be protection against *H. tamarinus*.*

* Since the foregoing was written, Trum (1966) in a communication to the International Symposium on Primate Diseases at Sukhumi has produced statistical evidence of human illness caused by *Herpes T.* acquired from bites or licks of squirrel monkeys. Symptoms varied from skin pustules to fever with cerebral symptoms and recovery.

Herpes-virus simiae (*B-virus*) Great concern has been caused in recent years by the fatalities amongst animal attendants caused by B-virus (*Herpes-virus simiae*) acquired from imported monkeys (*Macaca mulatta*). This concern has arisen more from the severe and usually fatal symptoms in human patients than from the number of cases which have occurred. The small number of incidents due to this disease in spite of the vast numbers of rhesus monkeys imported for medical research shows that the risk of infection is not great, although it is one to be taken seriously. The risk is greatest where large colonies of monkeys are maintained and where the turnover is high. Under such circumstances, the risk of a single animal carrying active infection is obviously higher than when, as in established menageries, monkeys are only imported to maintain the collections at a certain level. Nevertheless, newly imported monkeys should be handled with the greatest care and in such a way that scientists or animal attendants are not bitten. Those who examine monkeys *post-mortem* should carefully examine the mouth, lips and tongue for lesions before proceeding with their examination.

Herpes-virus simiae is closely related to the benign skin virus *H. simplex* of man, and indeed the two viruses have probably radiated from a single ancestral form, common to both groups. *H. simiae* causes little more annoyance to simian primates than does *H. simplex* to man. Good accounts of the disease both in primates and man are given by Keeble (1960), Davidson and Hummeler (1960) and MacLeod *et al.* (1960), in Whitelock (*Care and Diseases of the Research Monkey*. Whereas the disease in simian primates is associated with lesions of the tongue, buccal cavity, and the muco-epithelial border of the lips, infected humans suffer an acute, ascending myelitis which is rapidly fatal. The classical lesions in monkeys are herpes-like ulcers at the sites mentioned, initially consisting of small vesicles which soon rupture to give the ulcer, which heals below a fibrinous necrotic scab. The monkeys suffer little inconvenience even when the ulceration is severe and continue to eat without apparent difficulty. Patients show little sign by their general appearance of being infected and suffer little rise of temperature. A slight mucopurulent nasal discharge is often present but disappears about the tenth day; there is often a concurrent conjunctivitis of varying severity.

While tragedies arising from B-virus infections acquired from simian sources have been confined to contact with Asian species of monkeys, no confidence can be felt that it will not prove to be a natural infection in other Old-World species. For instance, Kalter *et al.* (1964) showed that thirty-eight of eighty-one baboon sera contained antibodies to B-virus, and forty-three did not. Thirty-three of thirty-eight positives were mature adults, indicating that they had passed through an acquired immunising infection since birth. The question as to whether the infection was one of B-virus or of *Herpes-virus simplex* remains unresolved.

Disquiet has been intensified by the death of a distinguished British scientist during the summer of 1965, following the bite of an African monkey, *Cercopithecus aethiops*. The patient was bitten on 26 April 1965 and became ill on 24 June, dying on 14 July. Unfortunately, tests on both the patient and the monkey were made too late, and a fully established diagnosis could not be made. The results of tests given under, are taken from an account of this case by G. Sumner-Smith (Jan. 1966), *Laboratory Primate Newsletter*, 5. The vervet monkey had a high neutralisation titre for both *H. simplex* and *H. simiae* B. The human patient showed virtually no neutralisation of *H. simplex*, but a markedly higher titre for *H. simiae B*, which increased as the disease progressed. No virus was isolated from cerebral tissues, cerebrospinal fluid, spinal cord or throat swabs from the patient, and no virus was isolated from the throat swab of the vervet.

The other monkeys in the colony consisted of four *Cercopithecus talapoin*, eight *C. mitis*, nine *C. aethiops* and ten *Erythrocebus patas*. All the monkeys were African species, and there had never been *Macaca mulatta* or *M. irus* (*cynomolgus*) in the colony.

The symptoms of the disease, from which the human patient suffered, were consistent with a diagnosis of *H. simiae* infection, as were the tests made on his serum. The monkey itself showed no lesions in the mouth, but after the rather long interval this is hardly surprising. The verdict therefore is unproven, but the case on balance of probability must be regarded as one of the transmission of B-virus infection to a human patient by an African monkey. Whether the patient was unduly susceptible, because of his lack of resistance to *H. simplex* virus, is not known but this remains a possibility.

The story of B-virus reinforces what has been said earlier in this work, under the heading of familial relationships, about the special dangers which arise from certain pathogens towards related species. The virus may not be of high infectivity, and it may be that *H. simplex* infection immunity confers a degree of protection against *H. simiae*; nevertheless, such related species, which have been ecologically separated, can be most dangerous companions and, until a clean bill of health is given, must be treated and handled with the greatest circumspection and care.

I Cytomegalo-viruses

'Salivary-gland' or 'cytomegalic inclusion' disease has been shown to infect monkeys (Cowdry and Scott, 1935a, b) and chimpanzees (Vogel and Pinkerton, 1955). Fiennes (1966) has shown that parotitis and other lesions of soft organs, similar to those associated with cytomegalic-inclusion disease, frequently accompany some of the common diseases of the skeleton, found in South American simian species. Salivary-gland disease, although associated for the most part with enlargements of the parotid and other salivary glands, which contain inclusion bodies, is associated also with severe systemic effects, including myocarditis. The main lesion is, however, to be found in the kidneys, and deaths are usually associated with nephritis. The disease can affect man, causing nephritis and a high incidence of death in very young children. Experience with other animals, such as guinea-pigs, indicates that cytomegalo-viruses – apart from the danger to human infants – may be 'passengers', and that they are very group-specific. It may therefore be doubted whether simian strains can infect human beings. There is no evidence that infection has passed from one to another at any time.

J Pox viruses

Ruch makes only cursory reference to the pox group of diseases in simians, although he quotes Bleyer (1922) who reported that New-World monkeys (*Mycetes*, or *Alouatta*, and *Cebus*) had participated in an outbreak of smallpox in human beings living in a jungle community in southern Brazil. Of varicella (chickenpox) he makes no mention, although this was subsequently described in chimpanzees, orang-utan and gorilla by Heuschele (1960). There appear to be no other established instances in which simian primates have been involved with man in diseases of the pox group, although it is well known from numerous experimental studies that most simian primates are susceptible to variola (smallpox). There is, however, no evidence that simians can maintain this disease in a wild community, and their importance in maintaining the disease in reservoir form is, so far as is known, negligible.

As with so many different species of animals, however, there appears to be a specific pox disease of monkeys, well described by Prier and Sauer (1960). This is a virulent disease, which can affect many species of apes and monkeys, although there is no evidence that it has ever affected a human being. Obviously, many people looking after captive monkeys and apes have been in contact with infection, but it is impossible to know whether they have been at risk, because all may well have been vaccinated. It is established that vaccinia virus will protect monkeys against monkey pox (McConnell *et al.*, 1964). Prier and Sauer describe the course and pathology of this disease, together with virological studies, which establish monkey-pox virus as being distinct from human pox viruses.

Peters (1965) described a further outbreak of monkey pox at the Blijdorp Zoo in Holland, which had devastating effects on the primate colony. His report, which appears to be adequately established, showed that the pox was introduced by newly imported anteaters (*Myrmecophaga tridactyla*). This suggests the possibility of a jungle pox cycle, in which animals other than monkeys are concerned. A précis of Peters' paper, given at the *VIIth*

International Meeting of Zoo Veterinarians, held at Zürich, Switzerland, in 1965, is given under:

A serious outbreak of monkey pox occurred in the monkey house, apparently brought into the Zoo by newly imported anteaters (*Myrmecophaga tridactyla*), who were housed close to the orang-utans, and showed lesions identical with those of smallpox in man. Ten days after the disease had been reported in the anteaters, the first male orang-utan showed the same clinical picture as the anteaters: profuse rhinitis, flat, yellowish-white blisters on the mucosa of the buccal cavity, lips, and the inside of the thighs, palms of the hands and plantar surfaces of the feet. The ulcerating blisters burst and started bleeding a couple of days later and the animal died on the 5th day.

Pox virus was isolated immediately by the specialist at the Public Health Department in Utrecht, the anteaters were killed, and pox virus was also shown. Disinfection, isolation of in-contact animals, as well as treatment with a proprietary drug (methisazonum) were started at once, but in spite of these precautions the disease swept through the orang-utan population, affecting eleven animals, of which only four recovered. The clinical symptoms as well as the pathological findings in all the affected apes were alike (with the exception of the rhinitis, which occurred in some animals only), and virus was isolated from most internal organs of all the animals that had died. In most cases there seemed to be some secondary infection connected with *Staphylococcus aureus*, *Streptococcus* sp., *Klebsiella pneumoniae*, or *Esch. coli*, and the author came to the conclusion that some of the animals might have survived the pox, but not the double infection.

Of four chimpanzees affected, none died, and the only two gorillas, although showing symptoms of the disease, recovered very quickly, as did four Hamlyn's owlfaced monkeys (*Cercopithecus hamlyn*), and one cotton-headed tamarin (*Leontocebus oedipus*), whereas three squirrel monkeys (*Saimiri sciureus*) out of a group of several died, one having aborted the day before; the other two Saimiri, at *post-mortem*, showed lesions of endometritis and chronic nephritis respectively. One gibbon (*Hylobates lar*)—the only one in the monkey house—died after 10 days. None of the siamangs, capuchins, Mona monkeys or Hoest cercopitheques were affected.

Yaba Virus Bearcroft and Jamieson (1958) described a disease in a colony of *Macaca mulatta* housed in the open at Yaba, Nigeria. The disease affected twenty of thirty-five rhesus monkeys in a few weeks, and also one baboon (*Papio papio*). A virus was isolated and described by Niven *et al.* (1961). The virus causes tumour-like growths on the heads and limbs of the monkeys, which regress in four to six weeks. The virus also produced local skin nodules in one man who was accidentally infected in the laboratory.

K Miscellaneous viruses

A review of virus studies on the normal baboon has been made by Kalter (1965) in *The Baboon in Medical Research.* Amongst other agents isolated from baboon tissue is lymphocytic choriomeningitis, in addition to a number of other viruses whose significance is not known. There is no evidence for supposing that rubella is a natural pathogen of simian primates. Owing to the increased importance attached to infections with this virus, more information about this is desirable; experimentally, Habel (1942) transmitted infection to *Macaca mulatta*, which developed leukopenia and in some cases a light-pink rash.

Possibly, by searching the literature, one could discover many more virus diseases to which simian primates are susceptible; one might suggest other human infections which primates might possibly carry, and which might add to the dangers of human contact with them. Information is too scanty for comfort, and some of it is so alarming as to raise doubts as to whether apes or monkeys should ever be permitted as companions for young human beings. Obviously this contact can only be permitted under the strictest of expert precautions. The danger that some unknown disease may be carried by simians and that they may disseminate it, unsuspected, over a considerable period is highlighted by the story of infectious hepatitis which, apart from the transmissible tumour viruses, is the last disease of viral origin to be considered in this work.

L Infectious hepatitis

Concern has been felt in the United States for some time about the high incidence of cases of infectious hepatitis amongst

HUMAN HEPATITIS WITH SUSPECTED SUB-HUMAN PRIMATE SOURCE
(Reports obtained by the Communicable Disease Center)

Year	State	Place	No. of suspected cases	Animals	Comments	Source
1953	Texas	Air Force base chimpanzee colony	5	Chimpanzees	Intimate contact with chimpanzees and monkeys	(1)
1955	Maryland	NIH animal colony	1	Chimpanzee	Veterinarian i/c of animal colony	(2)
1955	New York	City Zoo	2	Chimpanzees	Veterinarian and nurse treating sick zoo animals	(3)
1958	Florida	Monkey Jungle (exhibition)	1	Woolly monkey	Wife of owner: nursed sick baby monkey for 2 weeks, 3–4 weeks prior to onset	(4)
1958–60	Florida	Wild game farm and importation facility	4	Chimpanzees	Animal caretakers—contact with newly imported animals	(1)
1958–61	New Mexico	Air Force base chimpanzee colony	15	Chimpanzees	Veterinarians and animal caretakers contact with newly imported animals	(1) (5)
1960	Michigan	Private homes	8	Woolly monkey	Had been in 2 different homes approx. 6 weeks prior to onset of first case	(6)
1960	Michigan	Pet shop	1	Chimpanzee	Onset 6 weeks after caring for newly imported chimpanzee	(6)
1961	Florida	Center for Primate Research	8	Chimpanzees	Cases among persons working with newly imported young chimpanzees	(4) (7)
1961	New York	Animal importation facility	1	Chimpanzee	Animal caretaker—contact with newly imported animals	(8)
1961	Florida	Municipal Zoo	6	Chimpanzees	All cases had contact with newly imported baby chimpanzees	(4) (9)
1961	Connecticut	Home of Zoo Director	2	Chimpanzee	Wife and son of Zoo Director cared for newly imported chimpanzee 5 weeks before onset	(10)

INFECTIOUS HEPATITIS AMONG VARIOUS GROUPS AT HOLLOMAN AIR FORCE BASE, NEW MEXICO, AND SURROUNDING OTERO COUNTY MAY 1958 THROUGH DECEMBER 1960

Group	Population	Cases of Hepatitis	Incidence %
Otero County, New Mexico	36,870	71	0·2
Base personnel exclusive of Veterinary Services Branch	6,500	24	0·4
Handlers of chimpanzees in established colony only	75	0	0
Veterinary Services Branch	35	11	31
Intimate handlers of newly arrived chimpanzees (Veterinary Services Branch)	21	11	52

INFECTIOUS HEPATITIS AT YERKES LABORATORIES OF PRIMATE BIOLOGY

Date newly imported group of infant chimpanzees arrived at Laboratory	2 March 1961
Date of onset of first case	28 March 1961
Date of onset of last case	9 April 1961

Group	Number	Cases of Hepatitis
Persons having intimate contact with newly arrived chimpanzees	9	6
Persons having casual contact with newly arrived chimpanzees	12	2
Other employees having no contact with newly arrived chimpanzees	29	0
Totals	50	8

persons who have been in contact with newly imported chimpanzees and woolly monkeys. A summary of the cases to that date was included in the *Laboratory Primate Newsletter* (1962), from which some of the material here given, including the tables, is abstracted.* Hillis (1961), reported on the outbreaks at the Holloman Air Force Base in Florida, where a large number of chimpanzees are imported from West Africa for medical research. During a period of two and a half years the incidence of infectious hepatitis amongst those who came in intimate contact with the chimpanzees was 52% against 0·4% amongst other personnel employed there. The newly arrived chimpanzees were held in an isolation colony for forty-five days, before being admitted to the established colony. Amongst handlers of chimpanzees in the established colonies, there were no cases of infectious hepatitis during the two-and-a-half-year period. From this it was concluded that chimpanzees shed virus for no more than forty-five days, and for this reason it was supposed that infection was contracted from human sources while the chimpanzees were held in West African villages. It was not supposed that infectious hepatitis was a 'wild' disease of the chimpanzees. Nevertheless, the chimpanzees did not show gross symptoms of sufficient severity to attract attention.

Riopelle and Molloy (1963) also reported on outbreaks of infectious hepatitis amongst chimpanzee handlers at the Yerkes Laboratories in New Orleans. Six of nine persons having intimate contact with a newly imported group of baby chimpanzees suffered from hepatitis, in addition to two of twelve persons having only casual contact with them. The author describes virological, serological and pathological investigations into the disease and on infected chimpanzees. Liver biopsies on infected animals gave evidence of hepatitis, and careful clinical examination showed symptoms of jaundice in some cases. Liver-function tests also gave evidence of hepatic impairment.

Held (1962) is reported further in the *Laboratory Primate Newsletter* (Jan. 1964) summarising the position as known at that time. In this communication he stated that of the seventy-eight cases, chimpanzees were thought to have been the source of infection in sixty-one, woolly monkeys (*Lagothrix* sp.) in nine, a

* I am grateful to Dr Schrier for permission to reproduce these tables. Reference is also given to the original papers in which they appeared.

gorilla in four, and Celebes apes (*Cynopithecus niger*) in four. Owing to the importance of this disease, no apology is made for reproducing the tables contained in the *Laboratory Primate Newsletter* (1962).

Obviously a great deal more information is required with regard to this disease and to the range of simian primates which may transmit infection. The position revealed reinforces yet again the urgent need for further knowledge of the ecology of disease in simian primates, and of the pathogenesis of the diseases from which they may suffer, and of the ways in which they may endanger human beings.

M Transmissible Tumour Viruses as Zoonoses

Cancers in simian primates have been reviewed by Jungherr (1963); transmissible viruses of animals have been reviewed by Huebner (1963), from whom the table under is reproduced. Further studies were reported by Lapin (1966), Moloney (1966), Spencer-Munroe (1966) and Epstein *et al.* (1966) during the *Symposium on Comparative Medicine*, held June 1965 in London by the Zoological Society of London jointly with the World Health Organisation.

Huebner associates the transmissible cancers with four virus groups:

1 *The Papova group* The Papova group of viruses includes the infectious papillomas, as well as the SV-40 virus of monkeys (rhesus), which causes sarcoma in hamsters.
2 *The Adeno-viruses* Amongst the adeno-viruses, no. 12 and no. 18, derived both from man and from monkeys, can cause cancer in hamsters.*
3 *The Pox-viruses* Fibroma and myxoma viruses are the rabbit representatives of the pox group. One of these viruses causes monkey facial tumour, but has not been implicated in any human cancer. Huebner, however, quotes Duran-Reynals (1956) to the effect that vaccinia can (with oncogenic chemicals)

* Inconclusive results from the injection of adeno-virus 12 into immature baboons are reported by Kalter *et al.* (1966).

cause tumours. Yaba monkey virus (see above) is another member of the pox group which causes benign tumours in monkeys and man (Grace and Mirand, 1963).

4 *Leukaemia group* (*avian and murine*) None of these has so far been implicated in human and other primate tumours.

Simian primates are, therefore, undoubtedly at times carriers of viruses, which can prove oncogenic in other groups of animals, and possibly also man. Recent studies, e.g. Negroni (1964) and other workers also leave little doubt that some, at any rate, of the human leukaemias, and by implication the simian also, are of infectious origin. Studies of avian and murine infectious leukaemias suggest that the causal viruses may be responsible not only for the development of leukaemia, but also of other cancers, e.g. sarcoma as well.

Factual evidence is lacking as to whether simian oncogenic viruses – SV-40, adeno-viruses 12 and 18 – can pass cancers to humans as zoonoses. Similarly, more evidence is required as to the relationship of Burkitt's lymphoma to simian diseases in Old-World or New-World monkeys. It is in this area, that further research is required perhaps more urgently than in any other, and with possibly greater hopes of providing a major advance.

Four factors are of especial significance in the study of viruses as agents of oncogenic zoonoses:

1 the known ability of some viruses to cause inflammatory conditions in one host and cancers in others, e.g. SV-40, adenoviruses 12 and 18;
2 the ability of a single or closely related viruses to cause a variety of clinical conditions in different individuals of the same species, e.g. Rous sarcoma and the avian leucosis/petrosis group;
3 the ability of viruses to initiate malignant changes in cells, while subsequently defying demonstration by any laboratory means except the persistence of antibodies;
4 the ability of viruses to initiate malignant cellular reactions, if introduced with, or activated by, oncogenic agents.

There is, therefore, a serious danger that viruses from such

ANIMAL TUMOR VIRUSES

Virus	Natural host	Other hosts	Nucleic acid type	Viral inclusions	Comment
'Papova' group					
Polyoma	House mouse	Hamsters, rats, rabbits	DNA	Nucleus	K-virus of mice is probable member of this group
Shope papilloma	Cottontail rabbits	Domestic rabbits	,,	,,	
Bovine papilloma	Cattle	Horse, hamsters	,,	,,	
Human papilloma	Man	?	,,	,,	
Canine oral papilloma	Dogs	?	,,	,,	
SV-40 (vacuolating virus)	Monkeys (rhesus)	Man, hamsters, mastomys (concha rat)	,,	,,	
Adeno-virus group					
Type 12	Man	Hamsters develop cancers	,,	,,	Types 12 and 18 are unique in their absence of hemagglutinins
Type 18	,,	Hamsters develop cancers	,,	,,	
Pox-virus group					
Rabbit fibroma	Rabbit	?	,,	Cytoplasmic	Studies by Duran-Reynals suggest that vaccinia can (with oncogenic chemicals) cause tumours
Rabbit myxoma	,,	?	,,	,,	
Squirrel fibroma	Squirrel	Rabbit, woodchuck	,,	,,	
Monkey facial tumour	Monkey	?	,,	,,	
'Leukemia' group (mouse)					
Lymphocytic					
Gross	Probably house-mouse	Rat	RNA	?	1. The leukemia viruses have many of the properties of myxo-viruses
Moloney	,, ,, ,,	Rat, hamster	,,	Cytoplasmic	
Stem cell					
Friend	,, ,, ,,	,,	,,	,,	2. Purified human and bovine leukemia preparations show similar particles in the electron microscope
Myeloid					
Graffi	,, ,, ,,	,,	,,	,,	
Avian leukosis group					
Avian lymphomatosis	Chicken	Turkeys	,,	,,	
Avian erythroblastosis	,,	,,	,,	,,	
Avian myeloblastosis	,,	?	,,	,,	
Rous sarcoma	,,	Turkeys, hamsters, guinea pigs, rats, rabbits	,,	,,	
Mammary carcinoma	House mouse	None	,,	,,	
Kidney carcinoma	Frog	?	DNA ?		

closely related groups as the simian primates could show an altered pathogenesis in man, of which malignancy could be a feature. The dangers of such happening are enhanced by man's exposure in crowded cities to oncogenic agents and increased radiation hazards. It may prove that Burkitt's lymphoma is a zoonosis, and further information on this will be anxiously awaited. The situation in relation to this and to the oncogenic viruses is too confused to make further speculation worth while, although knowledge is likely to be quickly acquired.

SV-40, being a 'passenger' virus in monkey kidney cultures, must have been administered to thousands of humans in poliomyelitis vaccine; fortunately, no undesirable effects appear to have accrued, but these could be delayed for forty years or more. The danger of transmitting simian viruses in vaccines is a real and alarming one, and many will recall the tragedy – best forgotten – that accompanied the transmission of serum hepatitis in yellow-fever vaccine. In this case, however, the virus originated in human volunteers whose serum was used as a diluent.

Epstein *et al.* (1964) describe lymphomatous lesions in African green monkeys which had been injected two years previously, before weaning, with unfiltered material from a case of Burkitt's lymphoma in an African child. They referred also to material supplied by the author from South American monkeys (*Lagothrix* and *Cebus* spp.), from bone dyscrasias which were associated with metaplasia of the bone marrow and other lesions of lymphatous type present in internal organs. The syndrome associated with the bone dyscrasias in New World Primates has been described by Fiennes (1964, 1965), who believes that it is due to an infectious cause, possibly of viral origin. Whether the lesions of the skeleton are attributable, wholly or in part, to the infectious condition or whether they are due, wholly or in part, to nutritional and other metabolic causes is undecided. Epstein's suggestion that his green-monkey disease was associated with Burkitt's lymphoma was challenged by Wright and Bell (1964), who believed that the disease they described was of nutritional or metabolic origin, and who denied that the cellular reactions described could be those associated with the Burkitt tumour.

Epstein *et al.* (1966) gave a further account of work on the Burkitt tumour and reported that other monkeys injected with tumour material had again developed lesions of the skeleton, but that this time lymphomatous changes had been absent.

It is the general view that Epstein and his co-workers did not transmit the Burkitt tumour, and that the skeletal disease of New-World monkeys, which the author has associated with lymphomatous deposits, is likewise not of a malignant type. More need not be said, except that a lingering doubt remains, and that in a matter of such great importance this should be resolved by further study. Those with veterinary medical training associate the word lymphoma with the infectious lymphomas of birds, cats and cattle (see: Baker *et al.*; Dutcher *et al.*; Jarrett (1966)). The lymphomatous deposits described by the author and by Epstein and his co-workers in some respects resemble those of cats and cattle, and an association with primates in a disease cycle, in which birds play a part, could possibly occur. Thus, the Burkitt lymphoma could well be one of those diseases which arise from a mingling of arboreal and terrestrial cycles. Various workers have now described infections of young primates with Rous sarcoma virus, and important results may arise when all strains of the avian leukaemias have been tested in different primate species. A feature of the skeletal diseases of New-World primates is the involvement of the teeth in pathological changes, which would not occur in rickets or secondary hyperparathyroidism. It is significant that according to Merkalova (1965) deformities of bones and teeth have been shown to occur in hamsters infected with a latent rat virus. The association of osteopetrosis gallinarum with virus infection also shows that diseases of the bones can arise as a direct result.

It is perhaps appropriate to end our discussions on a note of speculation, perplexity and controversy. It must be obvious from what has been written how much is unknown, how much requires to be done, and what immense issues are involved in the closer contacts between man and simian primates, both ecologically and directly, which have arisen from the new ways of life which man has developed during the twentieth century.

10 Summary and Conclusions

In order to understand the zoonoses of primates, it is essential to study the parasites and diseases themselves, and also:

1 *the ecological relationships* between man and other primates, comprising direct and indirect contacts;

2 *the familial relationships* between them; these may be
   (a) actual, i.e. the two species may be *morphologically* closely related to each other, although
   (b) the point of divergence may, *in time*, be distant;

3 *the conditions which arise from phylogenetic propinquity and ecological separation*, related genera may have little direct contact with each other, and this is generally the case with man and other primates; man's direct contacts with other primates arise largely from artificial circumstances, created by him, except for certain situations with macaques in Asia.

With many monkey species, contacts between species are close; some occupy the upper forest canopies, others the middle and so on. This means that diseases and parasites are largely shared, and the different species are not dangerous to each other, because of the opportunities for the development of immunity to the parasites of alien groups occuping the same habitat.

In the case of man, most members of *Homo sapiens* occupy habitats, in which other primate species do not exist. In those areas, in which man dwells and other primates are also found, the habitats of each are very dissimilar and contacts are indirect. From this two important results emerge:

1 parasites and diseases of each group have radiated apart, as the groups themselves have evolved; this has taken place to such an extent that many parasites are not infectious to both groups; where they are infectious on the other hand, the dis-

eases caused are often of great severity in the unnatural group;
2 in order to understand these zoonoses, much more must be known of the life histories of the arthropods, which serve as intermediate hosts and vectors – mosquitoes, tsetse flies, *Chrysops* spp., ticks and many others that may prove to be important.

Simian primates, brought into captivity as specimens in menageries, as pets, or for medical research or other uses, are potentially dangerous to human beings unless they come from a known source or have been adequately quarantined and tested for pathogens. The danger to children is acute because simians frequently carry parasites or intestinal pathogens, particularly organisms of the dysentery sub-group of bacteria and Protozoa, such as *Amoeba*, *Giardia* and *Balantidium*.

Far too little is known of the virology of simians, but with regard to contacts with captive specimens, experience with *Herpes -virus simiae* and with infectious hepatitis in chimpanzees, should emphasize the dangers and the need for caution. The little that is known of the association of monkeys with potential oncogenic viruses emphasises that there may be new areas in which dangers could exist. The existence of so many 'passenger' viruses in primate faeces, potentially infectious or oncogenic, means that the greatest care should be taken in harvesting viruses for vaccine manufacture, where monkey cells are being used in tissue culture.

In general, New-World monkeys appear to be less dangerous to man than Old-World species. These could be more readily recommended as pets, especially for children. Unfortunately, the New-World monkeys tend to be delicate and are subject to a number of diseases which are as yet inadequately understood. If means could be found to overcome these diseases, it might be possible to recommend the New-World monkeys as safe and companionable pets; research on their diseases is justifiable on such grounds alone.

The dangers of free-living apes and monkeys to human populations, although contacts between the groups are only indirect, are obviously very great because the monkeys are exposed to, and are carriers of, so many wild arbo-viruses, to which the human species is unaccustomed and unduly susceptible. The two well-known examples of yellow fever and Kyasanur Forest disease illustrate both the dangers to human life and the possibility that fatal diseases may be spread from one endemic focus to create others continents apart. The illustration given of the supposed mutation of Chikungunya virus to O'nyong-nyong, with its distressing effects and rapid spread in man, illustrates the further dangers that simian viruses might become adapted to human populations, and spread with appalling rapidity, and under circumstances in which there were no possible immediate means of control.

Knowledge of prophylaxis against viral diseases is in its infancy, and time must elapse before any effective vaccine could be prepared, tested, and manufactured in bulk to protect populations against a pandemic caused by a new virus. Had O'nyong-nyong been attended by a high death-rate, comparable perhaps with that associated with myxomatosis when first introduced to a susceptible rabbit population, the human population of a large part of East and Central Africa would have virtually ceased to exist. To such an extent, in spite of twentieth-century medicine, is man still vulnerable to attack from new viruses.

Real advance in virology, on all fronts but particularly in the fields of prophylaxis and treatment, is thus of most urgent importance, and any future major tragedy may well have its source, as was found in the past, in the wild primate populations. Meanwhile, the 'wild' and 'passenger' viruses carried by simian primates require urgent study, so that at least, when some emergency occurs, their groups, characteristics, methods of propagation and attenuation may be understood. In this way, a new virus could be quickly modified for vaccine purposes, or related non-pathogenic viruses could be found to serve as such.

Plainly, it is in the realms of virology that primate zoonoses present the greatest danger. What of malaria? There is as yet no proof that human and simian malarias are being exchanged under natural conditions, although it is known that theoretically this could be so. This possibility must be considered in relation to malaria-eradication schemes, wherever human beings and simian primates share a malaria-infested area.

To the author, the preparation of a review on zoonoses of primates has proved a rewarding activity, but the greatest lesson learned has been that zoonoses are not just catalogues of diseases that can be passed – or are passed – between different groups of animals. Any student must acquire a knowledge of the familial and ecological conditions of the two groups and of the ways in which parasites can be passed from one to the other. These in the case of the primates are especially complex.

ADDENDA

(Rabies)

Since writing the foregoing on the subject of rabies, the first naturally occurring case of rabies has been diagnosed in a research laboratory of the Medical Research Council, Hampstead, London, England, in a rhesus monkey (*Macaca mulatta*). This was reported first in the London *Sunday Times* of 17 April 1966, and then by L. R. Boulger in the *Lancet* of 30 April 1966.

The case reported occurred in one animal from a group of rhesus monkeys newly imported from India on 15 November 1965. Fortunately, this monkey was in a cage by itself, although the other animals were caged in pairs.

Symptoms started to appear on the forty-seventh day, when the animal was noted to be off its food and cowering in the back of its cage. The animal, as monkeys often do when unfit, started to bite at its fingers and hands. By the fifty-first day the wounds so inflicted were of a severity necessitating the monkey's destruction. The animal had shown no aggression, no aversion to drinking, and there was no obvious paralysis.

Post-mortem it was found to be an adult female in a good state of nutrition. There were no lesions of the eyes, tongue, buccal cavity, or lips, and no apparent body scars. Fingers and hands, however, were grossly mutilated. The brain was congested, but the internal organs appeared normal. Material was taken from the brain, submandibular salivary glands and the trigeminal ganglia for further study. The existence of rabies was demonstrated by the presence of Negri bodies in the brain tissue, by passaging through mice and neutralisation by antirabies sera.

The monkey in question had shown no symptoms which could suggest that it was infected with rabies, and it was only by painstaking routine laboratory investigations that the existence of this disease was discovered.

This circumstance must give rise to speculation as to how many monkeys imported for research have contracted rabies in the past without diagnosis. Rabies must now take its place with B-virus and other infections as a deadly hazard, to which those handling newly imported monkeys may be exposed.

(Addendum to p. 128)

Dr W. C. Osman Hill (1966, pers. communication) has informed the writer of a case of suspected poliomyelitis in two gorillas, which occurred at the Yerkes Regional Primate Center about two years ago:

> Two gorillas were sick and showed symptoms of paresis of the hind limbs. One recovered but has been left with a permanent paralysis of at least one hind limb; he now gets about in an orthopaedic trolley and is doing very well. The other animal died, and his central nervous system was preserved and sent to a neuropathologist in Washington. A report was received on the sections, stating that lesions in the anterior horn cells of the spinal cord were characteristic of poliomyelitis. No lesions were seen in the cerebellum or the cortex. The possibility that the lesions had been caused by Coxsackie Virus had been ruled out.

Dr Hill stated that a colleague of his was working on the publication of these case histories at this time.

Acknowledgements

The *Zoonoses of Primates* was originally intended as a monograph to be prepared for distribution by the Food and Agriculture Organisation. I was first approached by Dr Ross Cockrill on behalf of FAO as long ago as July 1961. At the time the request to prepare a monograph seemed a simple one and I readily agreed. Once started, it quickly became apparent that this was by no means an easy task, and I was convinced that, in order to fulfil it satisfactorily, I must write about areas of knowledge that were not altogether familiar to me, namely the ecology and familial relationships of primates including man.

I had prepared what I believed was a suitable script for FAO purposes by June 1965, and about 100 copies of this were distributed gratis at the Zoological Society/World Health Organisation Symposium held at the London Zoo during that month.

Meanwhile, so much time and effort had been expended on the work that, following the advice of a number of persons who saw this script, it was decided to expand the material into full book form. Meanwhile, FAO reported that there would inevitably be some delay over publication and it was mutually agreed that the work should be offered to Messrs. Weidenfeld and Nicolson for commercial production.

I wish now to thank Dr Ross Cockrill for the original suggestion, which has I think proved a very rewarding one, and Dr Ervin A. Eichhorn, who has sponsored the work more recently, for constant interest and encouragement. The *Zoonoses of Primates*, which I think is much needed, has come into being as a result of the initiative of the Food and Agriculture Organisation, and I wish to give full acknowledgement to this.

Various sections of the work owe much to certain individuals who have given me guidance in areas in which they were expert. Over general zoology I have been greatly assisted by Dr L. Harrison Matthews, until recently Scientific Director to the Zoological Society of London; on primatology, I have been assisted by Dr John Napier, who has also contributed the classification table of primates and the distribution map. Dr Wright, British Museum (Natural History) has helped me over the helminthiasis section; and T. A. M. Nash, Esq., late Head of the West African Institute of Trypanosomiasis Research, on the section on tsetse-flies. Dr H. Trapido, Rockefeller Institute, has assisted me over the section on ticks, and on yellow fever and Kyasanur forest disease. On protozoology in general, and malaria in particular, I have had the advantage of assistance from two recognised experts in Dr C. A. Hoare, Wellcome Institute, and Professor J. C. C. Garnham, of the School of Hygiene and Tropical Medicine. On specialised areas of virology, I have been guided by Dr. A. J. Haddow, until recently Head of the Virus Research Institute, Entebbe, Uganda. I have also received the most useful help and criticism from Sir Christopher Andrewes, without whose recent book on *Viruses of Vertebrates* I would have had great difficulty in satisfactorily completing this section of the work.

The work as a whole has been read by Dr Hamish Innes, a leading expert on primate pathology, now working in USA at a laboratory with a 'thru-put' of 3,000 rhesus monkeys a year and a large breeding centre; I have to thank my old friend for putting me to the trouble of entirely rewriting certain sections of the book including that on tuberculosis. Dr J. W. Landells also read through and criticised the work in its entirety on behalf of the publishers and, in spite of all the care I had taken on it, found a number of 'schoolboy howlers' which I have gratefully corrected. Dr L. G. Goodwin, now Director of Science, Zoological Society, has also read the book as a whole and has contributed some useful suggestions on the helminthological material.

If I have forgotten to include any who have given me expert advice, I hope they will forgive me, because this is wholly unintentional. I shall not, in any case, forget to pay high tribute to those ladies who have not only typed the work but checked and rechecked the text and bibliography so carefully that I am hopeful that errors will be minimal. First and foremost, I express my thanks to my wife, Alice Isobel Fiennes, who has typed large sections of the work and helped me construct the index. My former secretary, Miss Moira Kelham, typed the early draft of the work and constructed the original bibliography; her successor Mrs Ethelyn Hazell ably followed her lead. Finally, I must thank Miss Erica von Bernuth who has typed the final drafts, checked the material and constructed the final draft of the bibliography.

The index has been constructed by myself and my wife. This has deliberately been made rather more full than is perhaps usual, in the hope that the work may, by its use, be of some value as a primate pathology as well as simply a monograph on Zoonoses. If the index is inadequate in any way, I have only myself to thank!

Finally I have included at the end a list of textbooks and some other works consulted, in order to give acknowledgement and recognition to the part that these have played in enabling this work to be compiled. The monograph by Stiles and Hassell on the parasites of primates, and the truly classical work of Theodore Ruch have been the foundation stones on which this edifice has been built. Ruch's work, in particular, leads an author readily to the requisite literature and even, I regret to say, tempts him to neglect direct reference to the original papers he quotes; in most cases I have made reference to the originals, but have found Ruch's quotations so apt that this has usually been of little avail.

The publishers have been in every way co-operative and helpful and, as usual, their Scientific Editor, Richard Carrington, has been a source of inspiration.

Bibliography

ABRAMOVA, YE. V. (1949): Spontaneous Dysentery in Monkeys: *Works of the Sukhumi Biological Station AMS USSR*, pp. 244–257.

——, (1952): Spontaneous Dysentery in Monkeys: *Ibid.*, Dissertation, 12 pp.

AJELLO, L. (1953): The Dermatophyte *Microsporum gypseum* as a Saprophyte and Parasite: *J. Invest. Derm.*, **21**, pp. 157–171.

—— (1954): Occurrence of *Histoplasma capsulatum* and other Pathogenic Moulds in Panamanian Soil: *Am. J. trop. Med. Hyg.*, **3**, 987.

AKSENOVA, A. S. (1958a): Newcastle Dysentery Bacilli Carriage in Monkeys: in *Voprosy infeksionnoy patologii v eksperimente na obezyanakh*, pp. 21–24. Sukhumi.

—— (1958b): Paratyphoid Bacilli Carriage in Monkeys: *Ibid.*, pp. 159–165. Sukhumi.

AKSENOVA, A. S. and KHUTSISHVILI, L. M. (1949): Paratyphoid Infection of Breslau Type in Monkeys: *Works of the Sukhumi Biological Station AMS USSR*, pp. 270–287.

AMBERSON, J. M. and SCHWARZ, E. (1952): *Ternidens deminutus* Railliet & Henry, a Nematode Parasite of Man and Primates: *Ann. trop. Med. Parasit.*, **46**, pp. 227–237.

ANAN'IN, V. V. and KARASEVA, YE. V. (1955): Leptospirosis in Monkeys: *Voprosy parazitologii*, **9**, pp. 96–98.

ANDREWES, SIR CHRISTOPHER (1964): *Viruses of Vertebrates*: London, Baillière, Tindall & Cox.

ANONYM (1961): Simian Malaria: *W.H.O. Chronicle*, **15**, pp. 23.

APPLEBY, E. C., GRAHAM-JONES, O. and KEEBLE, S. A. (1963): Primate Diseases Infectious to Man: *Vet. Rec.*, **75**, pp. 81–86.

BABLET, J., DESCHIENS, R. and PICK, F. (1951): Sur un cas de distomatose hépatique chez un papion: *Bull. Soc. Path. exot.*, **44**, pp. 297–298.

BACH, F. W., FUELSCHER, J. and HARNACK*, (1931): Ueber eine durch Affen verursachte Ruhrepidemie: *Veroeff. Med. Verw.*, **34**, 61–73

BAKER, C. G., CARRESE, L. M. and RAUSCHER, F. Jr. (1966): The Special Virus-Leukemia Program of the National Cancer Institute: Scientific Aspects and Program Logic: in 'Some Recent Developments in Comparative Medicine' [R.N. T- W- Fiennes, ed.], pp. 259–278. *Symp. zool. Soc., Lond.* No. 17. London, Academic Press.

BANKS, N. (1901): A New Genus of Endoparasitic Acarians: *Geneesk. Tijdschr. Ned. Ind.*, **41**, pp. 334–336.

BEARCROFT, W. G. C. and JAMIESON, M. F. (1958): An Outbreak of Subcutaneous Tumours in Rhesus Monkeys: *Nature, Lond.*, **182**, pp. 195–196.

BENHAM, R. W. and KESTEN, B. (1932): Sporotrichosis. Its Transmission to Plants and Animals: *J. Infect. Dis.*, **50**, pp. 437–458.

BERTRAM, D. S., MCGREGOR, I. A. and MCFADZEAN, J. A. (1958): Mosquitoes of the Colony and Protectorate of the Gambia: *Trans. R. Soc. trop. Med. Hyg.*, **52**, pp. 135–151.

BEYE, H. K., GETZ, M. E., COATNEY, G. R., ELDER, H. A. and EYLES, D. E. (1961): Simian Malaria in Man: *Am. J. trop. Med. Hyg.*, **10**, p. 311.

BLANC, G. and WOODWARD, T. E. (1945): The Infection of *Pedicinus albidas*, the Maggot's Louse, on Typhus-carrying Monkeys (*Macacus sylvanus*): *A. J. trop. Med.*, **25**, pp. 33–34.

BLEYER, J. C. (1922): Ueber auftreten von Variola unter Affen der Genera *Mycetes* und *Cebus* bei Vordringen einer Pockenepidemie in Urwaldgebieten an den Nebenfluessen des Alto Uruguay in Suedbrasilien: *Muench. med. Wochschr.*, **69**, p. 1009.

BOEHM, L. K. and SUPPERER, R. (1955): Zwei neue Lungenmilben aus Menschenaffen, *Pneumonyssus oudemansi* und *Pneumonyssus vitzthumi* (Acarina: Halarachnidae): *Oesterr. zool. Z.*, **6**, pp. 11–29.

BOULGER, L. R. (1966): Natural Rabies in a Laboratory Monkey: *Lancet*, **i**, pp. 941–943.

BRAY, R. S. (1956): Studies on Malaria in Chimpanzees. I. The Erythrocytic Forms of *Plasmodium reichenowi: J. Parasit.*, **42**, p. 588.

BRUG, S. L. and HAGA, J. (1930): Notes on the Sarcoptes Found in a Case of Scabies Crustosa and in a Case of Scabies in a Monkey: *Meded. Dienst. Dienst. Volksgezondh. Ned.-Ind.*, **19**, pp. 220–226.

BUCKLEY, J. J. C. (1960): On *Brugia* Gen. Nov. for *Wuchereria* Spp. of the 'Malayi' Group: *Ann. trop., Med. Parasit.*, **54**, pp. 75–77.

BUGHER, J. C., MANRIQUE, J. P., GARCIA, M. R. and MESSA, E. C. (1944): Epidemiology of Jungle Yellow Fever in Eastern Columbia: *Am. J. Hyg.*, **39**, pp. 16–51.

BURNET, SIR MACFARLANE (1962): *Natural History of Infectious Disease*: London, Cambridge University Press.

CABALLERO, E. and BARRERA, C. (1958): Estudios helminthologicos de la región onchocercosa de Mexico y de la República de Guatemala: *Nematoda*, **2**: '*Filarioidea*'.

CAMERON, T. W. M. (1928): A New Definite Host for *Schistosoma mansoni* (*Cercopithecus sabaeus*): *J. Helminth.*, **6**, pp. 219–222.

CARMICHAEL, J. and FIENNES, R. N. T- W- (1942): Rickettsia Infection of Dogs: *Vet. Rec.*, **54**, pp. 3–4.

CARPENTER, K. PATRICIA and COOK, E. R. N. (1965): An Attempt to Find Shigellae in Wild Primates: *J. comp. Path.*, **75**, pp. 201–204.

—— and SANDIFORD, B. R. (1952): Epidemiology of a Human Case Bacillary Dysentery Due to Infection by *Shigella flexneri* 103Z: *Br. med. J.* **i**, pp. 142–143.

CARRINGTON, R. (1963): *A Million Years of Man*: London, Weidenfeld & Nicolson.

CARTER, H. F., WEDD, G. and D'ABRERA, V.St.E. (1944); The Occurrence of Mites (Acarina) in Human Sputum and their Possible Significance: *Ind. med. Gaz.*, **79**, pp. 163–168.

CARTER, V. (1877): Spirillum Fever [as quoted by Castellani & Chalmers, A. J. in *Manual of Tropical Medicine*, 1913, 2nd ed.].

CASTELLANI and CHALMERS, A. J. (1913): *Pinoyella simiae*, p. 1023 [in *Manual of Tropical Medicine*, 2nd ed., London, Baillière, Tindall and Cox].

CATANEI, A. (1925): Présence d'un *Monilia* sur la langue des singes d'Algérie: *C.R. Soc. biol. Paris*, **93**, pp. 92–94.

CHANDLER, A. C. (1955): *Introduction to Parasitology with Special Reference to the Parasites of Man* [9th ed.]: New York, John Wiley & Sons.

CLARK, H. C. (1931): Progress in the Survey for Blood Parasites of the Wild Monkeys of Panama: *Am. J. trop. Med.*, **11**, pp. 11–20.

COATNEY, G. R., ELDER, H. A., CONTACOS, P. G., GETZ, M. E., GREENLAND, R., ROSSAN, R. N. and SCHMIDT, L. H. (1961): Transmission of the M. Strain of *Plasmodium cynomolgi* to Man: *Am. J. trop. Med. Hyg.* **10**, pp. 673–678.

COCKBURN, A. (1963): *The Evolution and Eradication of Infectious Diseases* Baltimore, Johns Hopkins.

CONANT, N. F. (1937): Studies in the Genus Microsporum: Taxonomic Studies: *Arch. Derm. Syph. Chicago*, **36**, pp. 781–808.

CONTACOS, P. G., ELDER, H. A. and COATNEY, G. R. (1962): Man to Man Transfer of two Strains of *Plasmodium cynomolgi* by Mosquito Bite: *Am. J. trop. Med. Hyg.*, **11**, pp. 186.

COON, CARLETON, S. (1962): *The Origin of Races*: London, Cope.

COURTOIS, G., SEGRETAIN, G., MARIOT, F. and LEVADITI, J. C. (1955): Mycose cutanée à corps leveriformes observée chez des singes africains en captivité: *Ann. Inst. Pasteur*, **89**, pp. 124–127.

COWDRY, E. V. and SCOTT, G. H. (1935a): Nuclear Inclusions Suggestive of Virus Action in Salivary Glands of the Monkey, *Cebus fatuellus*, I: *Proc. Soc. exp. Biol., N.Y.*, **32**, pp. 709–711.

—— (1935b): Nuclear Inclusions Suggestive of Virus Action in the Salivary Glands of the Monkey, *Cebus fatuellus*: *Amer. J. Path.*, **11**, pp. 647–658.

CUMMINGS, B. F. (1916): Studies on the Anoplura and Mallophaga, Being a Report upon a Collection from the Mammals and Birds in the Society's Gardens: *Proc. zool. Soc. Lond.*, **1**, pp. 253–295.

Davidson, W. L. and Hummeler, K. (1960): B Virus Infection in Man: *Ann. N.Y. Acad. Sci.*, **85**, p. 970.

Davis, L. J. (1945): Pulmonary Acariasis in Monkeys: *Br. med. J.*, **i**, p. 482.

Deinhardt, F. and Deinhardt, J. (1966): The Use of Platyrrhine Monkeys in Medical Research. In 'Some Recent Developments in Comparative Medicine' [R.N. T-W- Fiennes, ed.], pp. 127–152. *Symp. zool. Soc. Lond.*, No. 17: London, Academic Press.

Delorme, M. (1926): Transmission expérimentale de *Sarcoptes scabiei*, var. *cuniculi*, au cynocéphale (*Papio sphinx*, E. Geoff.): *Bull. Soc. Path. éxot.*, **19**, p. 899.

Deschiens, R. (1940): Relation de quatre cas d'infestation par *Watsonius watsoni* chez le papion: *Bull. Soc. Path. exot.*, **33**, pp. 396–400.

Desportes, C. and Roth, P. (1943): Helminthes récoltés au cours d'autopsies pratiquées sur différents mammifères morts à la ménagerie du Muséum de Paris: *Bull. Mus. Hist. nat. Paris*, **15**, pp. 108–114.

Dobell, C. (1936): Researches on the Intestinal Protozoa of Monkeys and Man. VIII. An Experimental Study of Some Simian Strains of *Entamoeba coli: Parasitology*, **28**, pp. 541–593.

Dodge, C. W. (1935): *Medical Mycology: Fungus Diseases of Man and other Mammals*: St. Louis, C. V. Moby.

Dorland, W. A. N. (1947): *The American Illustrated Medical Dictionary*: Philadelphia and London, Saunders.

Downs, W. G., Aitken, T. H. G. and Anderson, C. R. (1955): Activities of the Trinidad Regional Virus Laboratory in 1953 and 1954, with Special Reference to the Yellow Fever Outbreak in Trinidad, B.W.I.: *Am. J. trop. Med. Hyg.*, **4**, pp. 837–843.

Duke, B. O. L. (1954): The Transmission of Loiasis in the Forest Fringe Area of the British Cameroons: *Ann. trop. Med. Parasit.*, **48**, pp. 349–355.

—— (1955a): Studies on the Biting Habits of *Chrysops*: I. The Biting Cycle of *Chrysops silacea* at Various Heights above the Ground in the Rain-forest at Kumba, British Cameroons: *Ibid.*, **49**, pp. 193–202.

Duke, B. O. L. (1955b): Studies on the Biting Habits of *Chrysops*: II. The Effect of Wood Fires on the Biting Density of *Chrysops silacea* in the Rain-forest at Kumba, British Cameroons: *Ibid.*, **49**, pp. 260–272.

—— (1955c): Studies on the Biting Habits of *Chrysops*: III. The Effect of Groups of Persons, Stationary and Moving, on the Biting Density of *Chrysops silacea* at Ground Level in the Rain-forest at Kumba, British Cameroons: *Ibid.*, **49**, pp. 362–367.

—— (1955d): Studies on the Biting Habits of *Chrysops*: IV. The Dispersal of *Chrysops silacea* over Cleared Areas from the Rain-forest at Kumba, British Cameroons: *Ibid.*, **49**, pp. 368–375.

—— (1957): Experimental Transmission of *Loa loa* from Man to Monkey: *Nature, Lond.*, **179**, pp. 1357–1358.

DUKE, B. O. L. (1958): Studies on the Biting Habits of *Chrysops*: V. The Biting-Cycles and Infection Rates of *C. silacea, C. dimidiata, C. langi* and *C. centurionis* at Canopy Level in the Rain-forest at Bombe, British Cameroons: *Ann. trop. Med. Parasit.*, **52**, pp. 24–35.
—— (1959): Studies on the Biting Habits of *Chrysops:* VI. A Comparison of the Biting Habits, Monthly Biting Densities and Infection Rates of *C. silacea* and *C. dimidiata* (Bombe Form) in the Rain-forest at Kumba, Southern Cameroons, U.U.K.A.: *Ibid.*, **53**, pp. 203–214.
—— (1960): Studies on the Biting Habits of *Chrysops*: VII. The Biting-Cycles of Nulliparous and Parous *C. silacea* and *C. dimidiata* (Bombe Form): *Ibid.*, **54**, pp. 147–155.
—— (1962): Experimental Transmission of *Onchocerca volvulus* from Man to a Chimpanzee: *Trans. R. Soc. Med. Hyg.*, **56**, 271.
DUMAS, J. (1953): *Les animaux de laboratoire*: Paris, Editions médicales Flammarien.
DURAN-REYNALS, F. (1956): Realities and Hypotheses of Viral Infection as a Cause of Cancer: *Rev. canad. Biol.*, **14**, pp. 411–428.
DUTCHER, R. M., LARKIN, E. P., TUMILOWICZ, J. J., SZEKELY, I. E. and MARSHAK, R. R. (1966): Recent Researches on the Etiology of Bovine Lymphosarcoma: in 'Some Recent Developments in Comparative Medicine' [R. N. T-W- Fiennes, ed.], pp. 279–293. *Symp. zool. Soc. Lond.* No. 17: London, Academic Press.
DZHIKIDZE, E. K. (1954): Dysentery Carriers among Monkeys: *Zh. Mikrobiol. Epidemiol. Immunobiol.*, **4**, pp. 58–72.
—— (1956): *Shigella dysenteriae* Carriage in Monkeys: *Ibid.*, **10**, pp. 44–48.

EMMONS, C. W. (1940): *Trichophyton mentagrophytes* (*Pinoyella simii*) Isolated from Dermatophytosis in the Monkey: *Mycopath.*, **2**, pp. 317–319.
—— (1942): Isolation of *Coccidioides* from Soil and Rodents: *Publ. Hlth. Rep. Wash.*, **57**, pp. 109–111.
—— (1943): Coccidiodomycosis in Wild Rodents. A Method of Determining the Extent of Endemic Coccidioidomycosis: *Publ. Hlth. Rep. Wash.*, **58**, pp. 1–5.
—— (1947): *Biology of Coccidioides*. I. *Biology of Pathogenic Fungi*, pp. 71–82. London, Wm. Dawson & Sons Ltd.
—— (1949): Histoplasmosis in Animals: *Trans. N.Y. Acad. Sci.*, **2**, pp. 248–254.
—— (1951): Isolation of *Cryptococcus neoformans* from Soil: *J. Bact.*, **62**, pp. 685–690.
EPSTEIN, M. A., THOMSON, A. D. and WOODALL, J. P. (1966): Experiments with Burkitt's Lymphoma; Attempted Transmission to Monkeys in Relation to Virological Findings: in 'Some Recent Developments in Comparative Medicine' [R. N. T-W- Fiennes, ed.], pp. 323–335. *Symp. zool. Soc. Lond.*, No. 17, London, Academic Press.

EPSTEIN, M. A., WOODALL, J. P. and THOMSON, A. D. (1964): Lymphoblastic Lymphoma in Bone Marrow of African Green Monkeys (*Cercopithecus aethiops*) Inoculated with Biopsy Material from a Child with Burkitt's Lymphoma; *Lancet*, **ii**, pp. 288–291.

EYLES, D. E., COATNEY, G. R. and GETZ, M. E. (1960): *Vivax*-type Malaria Parasite of Macaques Transmissible to Man: *Science*, **131**, p. 1812.

FAWCETT, B. (1953): *Exploration Fawcett*: London, Hutchinson.

FENSTERMACHER, R. and JELLISON, W. L. (1932): Porcupine Louse Infesting the Monkey: *J. Parasitol.*, **18**, 294.

FIENNES, R. N. T-W- (1964a): *Man, Nature and Disease*: London, Weidenfeld and Nicolson.

—— (1964b): Cystic Bone Disease in South American Primates. II: *Proc. zool. Soc. Lond.*, **143**, pp. 521–523.

—— (1966a): South American Primate Disease: Soft Organ Lesions and their Significance, pp. 337–350.

—— (1966b): 'Some Recent Developments in Comparative Medicine', *Symp. zool. Soc. Lond.*, No. 17: London, Academic Press.

FONSECA, F. DA (1951): Plasmodio de primata do Brasil: *Mems. Inst. Oswaldo Cruz*, **49**, pp. 543.

FOX, H. (1923): *Disease in Captive Wild Mammals and Birds:* Philadelphia, Lippincott.

—— (1926a): Generalized ringworm in a young chimpanzee: *Rep. Lab. comp. Path. Philadelphia*, p. 30.

—— (1926b): Scabies in a Male Drill: *Rep. Lab. comp. Path. Philadelphia*, pp. 27–28.

FRIBOURG-BLANC, A., NIEL, G. and MOLLARET, H. H. (1963): Notes sur quelques aspects immunologiques du cynocéphale africain: *Bull. Soc. Path. exot.*, **56**, pp. 474–485.

FUERSTENBERG, M. H. F. (1861): *Die Kraetzmilben der Menschen und Tiere*: Leipzig, Wilhelm Engelmann.

FUHRMAN, D. P. (1954): A Revision of the Genus *Pneumonyssus* (Acarina: Halarachnidae): *J. Parasit.*, **40**, p. 31.

GALINDO, P., TRAPIDO, H. and CARPENTER, S. J. (1950): Observations on Diurnal Forest Mosquítoes in Relation to Sylvan Yellow Fever in Panama: *Amer. J. trop. Med.*, **30**, pp. 533–574.

GARNER, ANNA (1966): Private communication.

GARNHAM, P. C. C., HEISCH, R. B., MINTER, D. M., PHIPPS, J. D. and IKATA, M. (1961): *Culicoides adseri* Ingram and Macfie, 1923, a Presumed Vector of *Hepatocystis* (=*Plasmodium*) *kochi* (Laveran, 1899): *Nature, Lond.*, **190**, pp. 739–741.

GARNHAM, P. C. C., LAINSON, R. and GUNDERS, A. E. (1956): Some Observations on Malaria Parasites in a Chimpanzee: *Ann. Soc. belge Méd. trop.*, **36**, 811.

GIBSON, T. E. (1960): *Toxocara canis* as a Hazard to Public Health: *Vet. Rec.*, **72**, pp. 772–774.

GOODWIN, L. G. (1966): Personal communication.

GRACE, J. T. Jr. and MIRAND, E. A. (1963): Human Susceptibility to a Simian Tumor Virus: *Ann. N.Y. Acad. Sci.*, **108**, pp. 1123–1128.

GRAHAM, G. L. (1960): Parasitism in Monkeys: *Ibid.*, **85**, pp. 842–860.

GRZIMEK, B. (1951): Tod durch Lungenmilben bei einem Schimpansen: *Zool. Gart. Lpz.*, **18**, p. 249.

HABEL, K. (1942): Transmission of Rubella to *Macacus mulatta* Monkeys: *Publ. Hlth. Rep., Wash.*, **57**, pp. 1126–1139.

HACKETT, C. J. (1966): Personal communication.

HADDOW, A. J. (1961a): Studies on the Biting Habits and Medical Importance of East African Mosquitoes in the Genus Aïdes, II: Sub-genera *Mucidus diceromyia, finlayia* and *stegomya: Bull. ent. Res.*, **52**, pp. 317–351.

—— (1961b): Entomological Studies from a High Tower in Mpanga Forest, Uganda. VII. The Biting Behaviour of Mosquitoes and Tabanids: *Trans. R. ent. Soc. Lond.*, **113**, pp. 315–335.

—— (1964): Personal communication.

HADDOW, A. J., WILLIAMS, M. C., WOODALL, J. P., SIMPSON, D. I. H. and GOMA, L. K. H.: *Bull. Wld. Hlth. Org.* [in press].

HALLORAN, P. O'CONNOR (1955): A Bibliography of References to Diseases in Wild Mammals and Birds: *Amer. J. vet. Res.*, **16**, pp. 1–465.

HAMERTON, A. E. (1929]: Report on the Deaths Occurring in the Society's Gardens during 1928: *Proc. zool. Soc. Lond.*, 1929: p. 49.

—— (1934): Report on the Deaths Occurring in the Society's Gardens during 1933: *Ibid.*, 1934, pp. 389–422.

—— (1937): Report on the Deaths Occurring in the Society's Gardens during 1936: *Ibid.*, **107**, p. 443.

—— (1939): Pulmonary Acariasis in Monkeys: *4th Congr. Int. Path. Comp. Rome* (Atti e Comm.), p. 271.

HAMERTON, J. L. and KLINGER, H. P. (1963): Chromosomes and the Evolution of Man: *New Scientist*, **18**, p. 483.

HARRIS, R. J. C. (1962): *The Problems of Laboratory Animal Disease*: London, Academic Press.

HEISCH, R. B., GRAINGER, W. E., HARVEY, A. E. C. and LISTER, G. (1962): Feral Aspects of Rickettsial Infections in Kenya: *Trans. R. Soc. trop. Med. Hyg.*, **56**, pp. 272–282.

HELD, J. (1962): Hepatitis in Humans Associated with Chimpanzees: *U.S. Publ. Hlth. Serv., CDC Vet. Publ. Hlth. Noes, No. 8* [in *Laboratory Primate Newsletter*, 1, pp. 8–14].

HELWIG, F. C. (1925): Arachnid Infection in Monkeys (*Pneumonyssus foxi* of Weidman): *Am. J. Path.*, **1**, p. 389.

HENDERSON'S (1963): *Dictionary of Scientific Terms* [J. H. Kenneth, ed.]: London, Edinburgh; Oliver & Boyd.

HEUSCHELE, W. P. (1960): Varicella (Chickenpox) in Three Young Anthropoid Apes: *J. Am. vet. med. Ass.*, **136**, pp. 256–257.

—— (1961): Internal Parasitism of Monkeys with the Pentastomid, *Armillifer armillatus*: *Ibid.*, **139**, pp. 911–912.

HILL, W. C. O. [cited by Ruch (1959), p. 72, Philadelphia, W. B. Saunders].

HILLIS, W. D. (1961): Human Hepatitis with Suspected Sub-human Primate Source: *Am. J. Hyg.*, **73**, pp. 316–328.

HIRST, S. (1921): On Some New Parasitic Mites: *Proc. zool. Soc. Lond.*, 1921, pp. 769–802.

HOARE, C. A. (1962): Reservoir Hosts and Natural Foci of Human Protozoal Infections: *Acta trop.*, **19**, p. 281.

—— (1963): Does Chagas' Disease Exist in Asia?: *J. trop. Med. Hyg.*, **66**, pp. 297–298.

—— (1965): Personal communication.

HOLMES, A. W., CALDWELL, R. G., DEDMON, R. E. and DEINHARDT, F. (1964): Isolation and Characterization of a New Herpes Virus: *J. Immun.*, **92**, pp. 602–610.

HOOGSTRAAL, H. (1956): African Ixodoidea. I. Ticks of the Sudan (with Special Reference to Equatoria Province and with Preliminary Reviews of the Genera *Boophilus*, *Margaropus*, and *Hyalomma*. Research Report NM 005 050.29.27. Cairo, Egypt: *U.S. Naval Med. Res. Unit*, **3**, 1101 pp.

HOPKINS, G. H. E. (1949): The Host-Associations of the Lice of Mammals: *Proc. zool. Soc. Lond.*, **119**, p. 387.

HUEBNER, R. J. (1963): Tumor Virus Study Systems: *Ann. N.Y. Acad. Sci.*, **108**, p. 1129.

HULL, R. N., MINNER, J. R. and MASCOLI, C. C. (1958): New Viral Agents Recovered from Tissue Cultures of Monkey Kidney Cells. III. Recovery of Additional Agents Both from Cultures of Monkey Tissues and Directly from Tissues and Excreta: *Am. J. Hyg.*, **68**, pp. 31–44.

——, MINNER, J. R. and SMITH, J. W. (1956): New Viral Agents Recovered from Tissue Cultures of Monkey Kidney Cells. I. Origin and Properties of Cytopathogenic Agents S.V. 1, S.V. 2, S.V. 4, S.V. 6, S.V. 11, S.V. 12 and S.V. 15: *Ibid.*, **63**, pp. 204–215.

HUNT, R. D. and MELENDEZ, L. V. (1966): Spontaneous Herpes-T Infection in the Owl Monkey (*Aotus trivirgatus*): *Path. Vet.*, **3**, pp. 1–26.

INNES, J. R. M., COLTON, M. W., YEVICH, P. O. and SMITH, C. L. (1954): Lung Mites. Pulmonary Acariasis as an Enzootic Disease Caused by *Pneumonyssus simicola* in Imported Monkeys: *Am. J. Path.*, **30**, pp. 831–831.

—— (1965): Personal communication.

IVANOV, M. F. and AKSENOVA, A. S. (1961): Pathological Anatomy and Bacteriology of Pneumonia in Monkeys: in *Materialy konferentsii po biologii i patologii obezyan, Tezisy dkoladov*, p. 68: Sukhumi.

JARRETT, W. F. H. (1966): Recent Research in Leukaemia in the Cat: in 'Some Recent Developments in Comparative Medicine' [R. N. T-W-Fiennes, ed.], pp. 295–301. *Symp. zool. Soc. Lond.*, No. 17, London, Academic Press.

JONES, S. B. (1932): Intestinal Bilharziasis in St. Kitts, B.W.I.: *J. trop. Med. Hyg.*, **35**, pp. 129–136.

JOYEUX, C. and BAER, J. G. (1949): L'hôte normal de *Raillietina* (*R*) *demerariensis* (Daniels, 1895) en Guyane hollandaise: *Acta trop.*, **6**, p. 141.

JUNGHERR, E. (1963): Tumors and Tumor-like Conditions in Monkeys: *Ann. N.Y. Acad. Sci.*, **108**, pp. 777–792.

KALTER, S. S. (1965): Virological Studies on the Normal Baboon (*Papio doguera*): in 'The Baboon in Medical Research' [H. Vagtborg, ed.], *Proceedings of the First Symposium on the Baboon and Its Use as an Experimental Animal*, 1963. San Antonio, University of Texas Press.

——, RATNER, I. A., BRITTON, H. A., VICE, T. E., EUGSTER, A. K. and RODRIGUEZ, A. R. (1966): Adenovirus 12 Inoculations of Immature Baboons: in 'Some Recent Developments in Comparative Medicine' [R. N. T-W- Fiennes, ed.], pp. 303–307. *Symp. zool. Soc. Lond.*, No. 17, London, Academic Press.

——, RODRIGUEZ, A. R. and RATNER, I. A. (1964): Neutralising Antibodies in Baboon Serum to Measles and B Virus (*Herpesvirus simiae*). Paper presented at the American Society of Microbiology, 64th Ann. Mtg., Washington.

KAPLAN, W. G. (1959): Occurrence of Black Piedra in Primate Pelts: *Trop. geogr. Med.*, **11**, pp. 115–126.

——, LUCILLE, K. and AJELLO, L. (1958): Recent Developments in Animal Ringworm and their Public Health Implications: *Ann. N.Y. Acad. Sci.*, **70**, p. 636.

——, HENDRICKS, S. L. and LEEPER, R. A. (1957): Isolation of *Microsporum distortum* from Animals in the United States: *J. infect. Dis.*, **28**, pp. 449–453.

KAVTARDZE, K. N. (1958): Icteric Leptospirosis of Monkeys: in *Problems of Infectious Pathology in Experiments with Monkeys*, p. 189. Sukhumi.

KEEBLE, S. A. (1960): B Virus Infection in Monkeys: *Ann. N.Y. Acad. Sci.*, **85**, pp. 960–969.

KESSEL, J. F. and HOLTZWART, F. (1935): Experimental Studies with *Torula* from a Knee Infection in Man: *Am. J. trop. Med.*, **15**, pp. 467–478.

KNOWLES, R. and DAS GUPTA, B. M. A. (1932): A Study of Monkey-malaria, and Its Experimental Transmission to Man: *Ind. med. Gaz.*, **67**, pp. 301–320.

KOBAYASHI, H. (1920): On Some Digenetic Trematodes in Japan: *Parasitol.*, **12.**, pp. 380–410.

KOEGEL, A. (1951): *Zoonosen (Anthropozoonosen)*: Basel, Ernst Reinhardt.

KONOVA, K. P. and PONOMAROVA, P. A. (1949): Study on the Duration of Immunity to Measles in Experimental Monkeys: *Byull. eksp. Biol. Med.*, **28**, pp. 67–73.

——, SHISHLYANNIKOVA, M. A., RYAZANTSEVA, N. YE. and STUPINA, Z. N. (1945): Experimental Measles in Monkeys: *Zh. Mikrobiol. Epidemiol. Immunoboli.*, **1–2**, pp. 27–32.

LAPIN, B. A. (1956): Mite Bronchitis in Monkeys: *Arkh. Patol.*, **18**, p. 57.

——, and YAKOVLEVA, L. A. (1963): *Comparative Pathology in Monkeys*: Springfield, Ill., Thomas.

——, YAKOVLEVA, L. A. and CHERKOVICH, G. M. (1966): Use of Non-human Primates in Medical Research, Especially in the Study of Cardiovascular Pathology and Oncology: in 'Some Recent Developments in Comparative Medicine' [R. N. T-W- Fiennes, ed.], pp. 195–212. *Symp. zool. Soc. Lond.* No. 17, London, Academic Press.

LEIPER, R. T. (1911): On the Frequent Occurrence of *Physaloptera mordens* as an Intestinal Parasite of Man in Tropical Africa: *J. trop. Med. Hyg.*, **14**, p. 209.

LEWIS, D. J., HUGHES, T. P. and MAHAFFY, A. F. (1942): Experimental Transmission of Yellow Fever by Three Common Species of Mosquitoes from the Anglo-Egyptian Sudan: *Ann. trop. Med. Parasit.*, **36**, pp. 34–48.

LOCHTE, T. (1937): Ueber das Vorkommen der Piedra beim Schimpansen und ueber die Beziehungen der tierischen Piedra zur menschlichen: *Arch. Derm. Syph., Wien*, **175**, pp. 107–113.

LOVELL, R. (1929): Discussion on Monkeys and Human Diseases: *Proc. R. Soc. Med.*, **22**, pp. 24–27.

—— (1929): The Isolation of *Bacterium morgani* from the Mammals, Birds and Reptiles at the Zoological Gardens: *J. Path. Bact.*, **32**, pp. 79–83.

LUMSDEN, W. H. R. and BUXTON, A. P. (1951): A Study of the Epidemiology of Yellow Fever in the West Nile District, Uganda: *Roy. Soc. trop. Med. Hyg.*, **45**, pp. 53–78.

LUND, E. and PETERSEN, K. B. (1959): Pneumococci in Monkeys: *Acta path. microbiol. scand.*, **45**, pp. 309–313.

MAIR, N. S. (1965): Sources and Serological Classification of 177 Strains of *Pasteurella pseudotuberculosis* Isolated in Great Britain: *J. Path. Bact.*, 90, pp. 275–278.

MALHERBE, H. and HARWIN, R. (1957): Seven Viruses Isolated from the Vervet Monkey. *Br. J. exp. Path.*, **38**, pp. 539–541.

MANSON-BAHR, P. H. (1960): *Manson's Tropical Diseases.* A Manual of the Diseases of Warm Climates: London, Cassell.

MAYNE, R. G. (1860): *An Expository Lexicon of the Terms Ancient and Modern in Medical and General Science:* London.

—— (1862), *Medical Vocabulary*: London.

MELENDEZ, L. V., HUNT, R. D., GARCIA, F. G. and TRUM, B. F. (1966): A Latent Herpes-T Infection in *Saimiri sciureus* (Squirrel Monkey): in 'Some Recent Developments in Comparative Medicine' [R. N. T-W-Fiennes, ed.], pp. 393–397. *Symp. zool. Soc. Lond.*, No. 17: London, Academic Press.

MERINO, C., DOAN, C. A., WOOLPERT, O., SCHWAB, J. L. and SASLAW, S. (1941): Reactions of Monkeys to Experimental Respiratory Infections. III. Response to Mixtures of Influenza Virus and Streptococcus: *Proc. Soc. exp. Biol. Med.*, **48**, p. 563.

MERKALOVA, Z. I. (1965): Deformity of Bones and Teeth in Hamsters Caused by Latent Rat Virus: *Vopr. Visusologii*, **10**, pp. 406–409.

MILLER, J. H. (1959): The Dog-face Baboon, *Papio doguera*, a Primate Reservoir Host of *Schistosoma mansoni* in East Africa: *J. Parasit*, **45**, p. 22.

—— (1960): *Papio doguera* (Dog-face Baboon) a Primate Reservoir Host of *Schistosoma mansoni* in East Africa: *Trans. R. Soc. trop. Med. Hyg.*, **54**, pp. 44–46.

MOLONEY, J. B. (1966): The Application of Studies in Murine Leukemia to the Problem of Human Neoplasia: in 'Some Recent Developments in Comparative Medicine' [R. N. T-W-Fiennes, ed.], pp. 251–258. *Symp. zool. Soc. Lond.*, No. 17: London, Academic Press.

MONBREUN, DE W. (1935): Experimental Cutaneous Blastomycosis in Monkeys: A Study of the Etiologic Agent: *Arch. Derm. Syph. N.Y.*, **31**, pp. 831–854.

MULZER, P. and NOTHHAAS, R. (1926): Ueber einen Fall von ausgedehnter Microsporie der Haut bei einem syphylitischen Affen: *Wien. med. Wschr.*, **76**, pp. 912–914.

MACARTHUR, W. P. (1924): A Case of Infestation of Human Liver with *Hepaticola hepatica* (Bancroft, 1893; Hall, 1916), with Sections from the Liver: *Proc. R. Soc. Med.*, **17**, pp. 83–84.

MCCONNELL, S., HERMAN, Y. F., MATTSON, D. E., HUXSOLL, D. L., LANG, C. M. and YAGER, R. H. (1964): Protection of Rhesus Monkeys against Monkeypox by Vaccinia Virus Immunization: *J. vet. Res.* **25**, pp. 192–195.

MCKENNEY, F. D., TRAUM, J. and BONESTELL, A. E. (1944): Acute Coccidioidomycosis in a Mountain Gorilla (*Gorilla beringeri*): *J. Am. vet. med. Ass.*, **104**, pp. 136–140.

MACLEOD, D. R. E., SHIMADA, F. T. and WALCROFT, M. J. (1960): Experimental Immunization against B Virus: *Ann. N.Y. Acad. Sci.*, **85** p. 980.

NAPIER, J. R. and NAPIER, P. H. (in press): *A Handbook of Living Primates*: London, Academic Press.

NASH, T. A. M. (1948): *Tsetse Flies in British West Africa*: HMSO, p. 66.

NEGRONI, G. (1964): Isolation of Viruses from Leukaemia Patients: *Br. med. J.*, i, pp. 927–929.

NEGRONI, P. (1965): *Histoplasmosis – Diagnosis and Treatment*: Springfield, Ill., Thomas.

NELSON, G. S. (1960): Schistosome Infections as Zoonoses in Africa: *Trans. R. Soc. trop. Med. Hyg.*, **10**, pp. 301–316.

—— (1965): The Parasitic Helminths of Baboons with Particular Reference to Species Transmissible to Man: in *The Baboon* [H. Vagtborg, ed.], pp. 441–470: Austin, University of Texas Press.

NEW YORK ACADEMY OF SCIENCE (1960): *Care and Maintenance of the Research Monkey* [O.v.St. Whitelock, ed.], **85**, pp. 735–992.

NICOLLE, C. and LUMBROSO, U. (1926): Research on the Granular Conjunctivitis of Some Laboratory Animals, especially the Rabbit. Their Importance in the Experimental Investigation of the Human Trachoma: *Archs. Inst. Pasteur Tunis*, **15**, pp. 240–261.

NIVEN, J. S. F., ARMSTRONG, J. A., ANDREWES, C. H., PEREIRA, H. G. and VALENTINE, R. G. (1961): Subcutaneous 'Growth' in Monkeys Produced by a Poxvirus: *J. Path. Bact.*, **81**, pp. 1–14.

NOUVEL, J. (1954): Spirochétoses des Animaux Sauvages: *Mammalia*, **18**, pp. 112–123.

——, BULLIER, P. and RIJNARD, J. (1957): Rapport sur la mortalité et la natalité enregistrées au Parc zoologique pendant l'année 1956: *Bull. Mus. Hist. nat. Paris*, **29**, p. 297.

—— and RIJNARD, J. (1949): Pseudotuberculose du singe cynocéphale (*Papio papio*) (Desm.) à bacille de Malassez et Vignal: *Rev. Path. comp.* **49**, pp. 68–70.

PETERS, J. C. (1965): Eine 'Monkey-Pox' Enzootie im Affenhaus des Tiergartens 'Blijdorp': *VII. Int. Symp. Diseases in Zoo Animals*, 1965, pp. 197–201.

PHILIPPE, J. (1948): Note sur les gales du singe. *Bull. Soc. Path. exot.*, **41**, p. 597.

PROBSTMAYER, W. (1963): *Dictionary of Veterinary Medicine*.

Pí, GEORGES SABATER (1963): Private communication.

PICK, F. (1951): Sur un nouveau trématode du genre *Watsonius* chez *Papio sphinx*: *Bull. Soc. Path. exot.*, **44**, pp. 59–61.

PILLERS, A. W. N. (1921): Sarcoptic Scabies (or Itch) in the Chimpanzee: *Vet. J.*, **77**, p. 329.

PINOY, E. (1912): Epidermophtye du singe: *Bull. Soc. Path. exot.*, 5, pp. 60–63.

POKORNY, J., HUEBNER, J. and ZÁSTĚRA, M. (1961): Isolation of *Toxoplasma gondii* from Certain Domestic and Free-living Animals in Czechoslovakia: *Cslká Epidem. Mikrobiol. Immunol.*, **10**, pp. 323–329.

POTTS, W. H. (1955): A New Tsetse Fly from the British Cameroons: *Ann. trop. Med. Parasit.*, **49**, pp. 218–226.

PRICE, E. W. (1928): New Helminth Parasites from Central American Mammals: *Proc. U.S. Nat. Mus.*, **73**, pp. 1–7.

PRIER, J. E. and SAUER, R. M. (1960): A Pox Disease of Monkeys: *Ann. N.Y. Acad. Sci.*, **85**, pp. 951–959.

RAILLIET, A., HENRY, A. C. L. and JOYEUX, C. E. (1912): Sur deux trématodes de primates: *Bull. Soc. Path. exot.*, **5**, pp. 833–837.

RATCLIFFE, H. L. (1954): Causes of Death in the Animal Collection: *Rep. Penrose Res. Lab.*, 1954, pp. 6–16.

—— (1955): Causes of Death in the Animal Collection: *Ibid.*, 1955, pp. 7–19.

—— and WORTH, C. B. (1951): Toxoplasmosis of Captive Wild Birds and Mammals: *Am. J. Path.*, **27**, 655–667.

REID, J. A. and WEITZ, B. (1961): Anopheline Mosquitoes as Vectors of Animal Malaria in Malaya: *Ann. trop. Med. Parasit.*, **55**, 180.

REIMAN, H. A. and KUROTCHKIN, T. O. (1931): Attempts to Produce Bronchomoniliasis in Monkeys: *Am. J. trop. Med.*, **11**, pp. 151–155.

REPORT OF INTERNATIONAL STUDY GROUPS (1963): *Virology*, **19**, p. 114.

RETNASABAPATHY, A. and JOSEPH, P. G. (1966): Melioidosis in a Monkey: *Vet. Rec.*, **79**, 72–73.

REWELL, R. E. (1948): Diseases of Tropical Origin in Captive Wild Animals: *Trans. R. Soc. trop. Med. Hyg.*, **42**, p. 17.

—— (1949): Outbreak of *Shigella schmitzi* Infection in Men and Apes: *Lancet*, **256**, pp. 220–221.

RIOPELLE, A. and MOLLOY, J. F. (1962): Infectious Hepatitis at Yerkes Laboratories of Primate Biology: *Laboratory Primate Newsletter*, **1** (4), p. 12.

ROBINSON, D. T., ARMSTRONG, E. and CARPENTER, P. (1965): Outbreak of Dysentery due to Contact with a Pet Monkey: *Br. med. J.*, **i**, pp. 903–905.

RODHAIN, J. (1939): Les plasmodiums des anthropoïdes de l'Afrique centrale et leurs relations avec les plasmodiums humains: *Annls. Soc. belge Méd. trop.*, **19**, p. 263.

—— and DELLAERT, R. (1955): Contribution à l'étude de *Plasmodium schwetzi* 1 et 2: *Annls. Soc. belge Méd. trop.*, **35**, 73; p. 757.

ROSS, R. W. and GILLETT, J. D. (1950): The Cyclical Transmission of Yellow Fever Virus through the *Grivet* Monkey, *Cercopithecus aethiops centralis* Neumann, and the Mosquito *Aëdes* (*Stegomyia*) *africanus* Theobald: *Ann. trop. Med. Parasit.*, **44**, pp. 351–356.

Roy, S. C. (1938): '*Bertiella studeri*, Natural Tape-Worm Parasite of Monkeys, in Hindu Child', *Ind. med. Gaz.*, **23**, 346.

Ruch, T. C. (1959): *Diseases of Laboratory Primates*: Philadelphia, W. B. Saunders.

Ryazantseva, N. Ye. (1956): Outbreak of Measles among Monkeys: *Zh. Mikrobiol. Epidemiol. Immunobiol.*, No. 4, p. 88.

Sabin, A. B. (1959): Reoviruses – A New Group of Respiratory and Enteric Viruses, formerly Classified as ECHO Type 10, is Described: *Science*, **130**, pp. 1387–1389.

Sanderson, I. T. (1957): *The Monkey Kingdom*: London, Hamish, Hamilton.

Sandground, J. H. (1936): On the Occurrence of a Species of *Loa* in Monkeys in the Belgian Congo: *Ann. Soc. belge Méd. trop.*, **16**, pp. 273–278.

Saslaw, S., Wilson, H. E., Doan, C. A., Woolpert, O. C. and Schwab, J. L. (1946): Reactions of Monkeys to Experimentally Induced Influenza Virus A Infection. An Analysis of the Relative Rôles of Humoral and Cellular Immunity under Conditions of Deficient Nutrition: *J. exp. Med.*, **84**, pp. 113–125.

Sauer, R. M. and Fegley, H. C. (1960): The Rôles of Infectious and Noninfectious Diseases in Monkey Health: *Ann. N. Y. Acad. Sci.*, **85**, pp. 866–888.

Schaeffer, M. (1962): Virus and Rickettsial Diseases – Methods and Criteria of Diagnosis: *Proc. Int. Symp. comp. Med.*, p. 134.

Schneider, N. J., Prather, E. C., Lewis, A. L., Scatterday, J. E. and Hardy, A. V. (1960): Enteric Bacteriological Findings on Monkeys Newly Received at Okatie Farms During 1955: *Ann. N. Y. Acad. Sci.*, **85**, pp. 935–941.

Schultze, C. H.: [in Mayne, R. G., 1860].

Scott, H. H. (1926): Report on the Deaths Occurring in the Society's Gardens During 1925: *Proc. zool. Soc. Lond.*, 1926, p. 231.

—— (1927): During 1926: *Ibid.*, 1927, p. 173.

—— (1927): Tuberculosis in Captive Wild Animals as Compared and Contrasted with the Disease in Man: *Proc. R. Soc. Med.*, **20**, p. 197–200.

—— (1928): Report on the Deaths Occurring in the Society's Gardens During 1927: *Proc. zool. Soc. Lond.*, 1928, p. 89.

—— (1930): Tuberculosis in Man and Captive Animals. A Study in Comparative Pathology. *Spec. Rept. Med. Res. Council*, No. 149, p. 270: London.

Shannon, R. C. and Greene, C. T. (1926): A Botfly Parasite in Monkeys: *Zoopathologica*, **1**, p. 285.

Shaw, G. I. (1959): Private Communication on Schistosomiasis on St. Kitts. [in: *Nelson*, G. S., 1960].

SIMPSON, J. J. (1911): Entomological Research in British West Africa. Part I, Gambia: *Bull. ent. Res.*, **2**, p. 187.

SMITHBURN, K. C. and HADDOW, A. J. (1946): Isolation of Yellow Fever Virus from African Mosquitoes: *Am. J. trop. Med.*, **26**, pp. 261–271.

—— and —— (1949): The Susceptibility of African Wild Animals to Yellow Fever, I. Monkeys: *Ibid.*, **29**, pp. 389–408.

SOYSA, E. and JAYARWADENA, M. S. D. (1945): Pulmonary Acariasis: a Possible Cause of Asthma: *Br. med. J.*, **i**, p. 1.

SPENCER-MUNROE, J. (1966): Viral Oncogenesis in the Rhesus Monkey: Miscellaneous Studies: in 'Some Recent Developments in Comparative Medicine' [R. N. T-W- Fiennes, ed.], pp. 229-250. *Symp. zool. Soc., Lond.*, No. 17, London, Academic Press.

SPITZKER, E. C. (1878): Sektionsberichte aus dem pathologisch-anatomischen Institut der New Yorker Thierarznei-Schule: *Zool. Gart. Frankfurt*, **19**, pp. 233–238.

SPRENT, J. F. A. (1956): The Life History and Development of *Toxocara cati* (Schrank, 1788) in the Domestic Cat: *Parasitology*, **46**, pp. 54–78.

—— (1958): Observations on the Development of *Toxocara canis* (Werner, 1782) in the Dog: *Ibid.*, **48**, pp. 184–209.

STILES, C. W. and HASSALL, A. (1929): Key-catalogue of Parasites Reported for Primates (Monkeys and Lemurs) with their Possible Public Health Importance: *U.S. Hyg. Lab. Bull.* (152), p. 409.

STRONG, J. P., MCGILL, H. C. and MILLER, J. H. (1961): Schistosomiasis mansoni in the Kenya Baboon: *Am. J. trop. Med. Hyg.*, **10**, pp. 25–31.

——, MILLER, J. H. and MCGILL, H. C. (1963): Naturally Occurring Parasitic and Other Lesions in Baboons: in *The Baboon in Medical Research* [H. Vagtborg, ed.]: Austin, University of Texas Press.

STUNKARD, H. W. (1923): On the Structure, Occurrence and Significance of *Athesmia foxi*, a Liver Fluke of American Monkeys: *J. Parasit.*, **10**, pp. 71–79.

—— (1949): Dicrocoelid Trematodes from the Gorilla: *Ibid.*, **35**, pp. 22–23.

—— and GOSS, L. J. (1950): *Eurytrema brumpti* Railliet, Henry & Joyeux, 1912 (Trematoda: Dicrocoeliidae) from the Pancreas and Liver of African Anthropoid Apes: *Ibid.*, **36**, pp. 574–581.

SUMNER-SMITH, G. (1966): B Virus in Association with a Monkey Colony at a Department of Psychology: *Laboratory Primate Newsletter*, **5** (1), pp. 1–4.

SUREAU, P., RAYNAUD, J. P., LAPEIRE, C. and BRYGOO, E. R. (1962): Premier isolement de *Toxoplasma gondii* à Madagascar. Toxoplasmose spontanée et expérimentale du *Lemur catta: Bull. Soc. Path. exot.*, **55**, pp. 357–362.

TAKOS, M. J. and ELTON, H. W. (1953): Spontaneous Cryptococcosis of Marmoset Monkeys in Panama: *Archs. Path.*, **55**, pp. 403–407.

TAMMEMAGI, L. and JOHNSTON, L. A. Y. (1963): Melioidosis in an Orang-outang in North Queensland: *Aust. vet. J.*, **39**, pp. 241–242.

TICTIN, I. (1897): Infection of Monkeys by Inoculation of the Blood Obtained by Crushing Bugs, which Had Recently Fed on a Patient: in *Manual of Tropical Medicine* [A. Castellani and A. J. Chalmers, 2nd ed.], p. 926, (1913), London, Baillière, Tindall & Co.

TRAPIDO, H. (1964): Personal communication.

——, GALINDO, P. and CARPENTER, S. J. (1955): A Survey of Forest Mosquitoes in Relation to Sylvan Yellow Fever in the Panama Isthmian Areas: *Am. J. trop. Med.*, **4**, pp. 525–542.

——, RAJAGOPLAN and WORK, T. H. (1959): Kyasanur Forest Disease, Part 8. Isolation of Kyasanur Forest Disease Virus from Naturally Infected Ticks of the Genus *Haemaphysalis: Ind. J. med. Rev.*, **47**, pp. 133–138.

——, and WORK, T. H. (1962): Non-human Vertebrates and Hosts and Disseminators of Kyasanur Forest Disease: *Proc. 9th Pacific Sci. Congr.*, **17**, pp. 85–87.

VANACE, P. W. (1960): Experimental Streptococcal Infection in the Rhesus Monkey: *Ann. N.Y. Acad. Sci.*, **85**, pp. 910–930.

VANBREUSEGHAM, R. (1958): *Mycoses of Man and Animals*: London, Sir Isaac Pitman and Sons, p. 235.

VITZTHUM, H. (1930): *Pneumonyssus stammeri*, ein neuer Lungenparasit: *Z. Parasitenk.*, **2**, 595–615.

VOGEL, F. S. and PINKERTON, H. (1955): Spontaneous Salivary Gland Virus Disease in Chimpanzees: *Archs. Path.*, **60**, pp. 281–285.

VORONIN, L. G., KANFOR, I. S., LAKIN, G. F. and TIKH, N. N. (1948): Spontaneous Diseases of Lower Monkeys, their Prophylaxis, Diagnosis, and Treatment: in *Experimentation on the Keeping and Raising of Monkeys at Sukhumi*, Chapt. 3, Moscow, Academy of Medical Sciences.

WALKER, A. E. (1936): *Cysticercosis cellulosae* in the Monkey: A Case Report: *J. comp. Path. Ther.*, **49**, pp. 141–145.

WALKER, J. and SPOONER, E. T. C. (1960): Natural Infection of the African Baboon *Papio papio* with the Large-cell Form of *Histoplasma*: *J. Path. Bact.*, **80**, pp. 436–438.

WARREN, M., EYLES, D. E., WHARTON, R. H. and KONG, O. Y. C. (1962): Susceptibility of Malayan Anophelines to Malaria Infection with *Plasmodium cynomolgi bastianelli: Trans. R. Soc. trop. Med. Hyg.*, **56**, p. 259.

WATSON, J. M. (1960): *Human Helminthology*: pp. 364–368. London, Baillière, Tindall and Cox.

WEIDMAN, F. D. (1923): Certain Dermatoses of Monkeys and an Ape. Pemphigus, Scabies, Sebaceous Cyst, Local Subcutaneous Edema, Benign Superficial Blastomycotic Dermatosis and *Tinea capitis* and *circinata*. *Arch. Derm. Syph. N.Y.*, **7**, pp. 289–302.

WEIDMAN, F. D. (1927): Observations on Skin Conditions: *Rep. Lab. comp. Path. Philadelphia*, pp. 36–38.

—— (1928): Zoo Tales – Being a Pathologist's: *The Centaur*, 1928, pp. 95–114.

—— (1933): Cutaneous Torulosis: The Identification of Yeast Cells in General in Histological Sections: *Sth. med. J. Nashville*, **26**, pp. 851–863.

—— (1935): Dermatoses of Monkeys and Apes: *9th Inter. Congr. Dermatol.*, 1935, Deliberations, **1**, pp. 600–606.

WEITZ, B. (1963): The Feeding Habits of Glossina: *Bull. Wld. Hlth. Org.*, **28**, pp. 711–729.

WHARTON, R. H. and EYLES, D. E. (1961): *Anopheles hackeri*, a Vector of *Plasmodium knowlesi* in Malaya: *Science*, **134**, p. 279.

WILSON, S. and THOMPSON, A. E. (1964): A Fatal Case of Strongyloidiasis: *J. Path. Bact.*, **87**, pp. 169–176.

WOMERSLEY, H. (1952): The Scrub-typhus and Scrub-itch Mites (*Trombiculidae*, Acarina) of the Asiatic-Pacific Region: *Rec. S. Aust. Mus.*, **10**, 673 pp.

WOOLPERT, O. C., SCHWAB, J. L., SASLAW, S., MERINO, C. and DOAN, C. A. (1941): Reactions of Monkeys to Experimental Respiratory Infections. I. Response to Influenza Virus A: *Proc. Soc. exp. Biol., N.Y.*, **48**, pp. 558–560.

WORLD HEALTH ORGANISATION (1959): WHO/FAO Expert Committee on Zoonoses: *Tech. Rept. Series*, No. 169.

WRIGHT, C. (1965): Private communication.

WRIGHT, D. H. and BELL, T. M. (1964): Bone Disease in African Green Monkeys: *Lancet*, **ii**, pp. 969–970.

YOUNG, R. J., FREMMING, B. D., BENSON, R. E. and HARRIS, M. D. (1957): Care and Management of a *Macaca mulatta* Monkey Colony: *Proc. Anim. Care Panel*, **7**, pp. 67–82.

ZUCKERMAN, S. (1933): *Functional Affinities of Men, Monkeys, and Apes:* London, Kegan Paul.

* Incredible as it may seem, no copy of this paper can be obtained, and all abstracts omit Harnack's initials, as does Ruch in his reference.

List of Reference Books

AINSWORTH, G. C. (1952): *Medical Mycology*: London, Pitman.

—— and AUSTWICK, P. K. C. (1959): *Fungal Diseases of Animals:* Commonwealth Agric. Bureau, Farnham Royal.

ANDREWES, SIR CHRISTOPHER (1964): *Viruses of Vertebrates:* London, Baillière, Tindall & Cox.

FOX, H. (1923): *Disease in Captive Wild Mammals and Birds:* Philadelphia, Lippincott.

GROSS, L. (1961): *Oncogenic Viruses:* Pergamon, New York, Oxford, London, Paris.

HADDOW, A. J. (1955–62): *East African Virus Research Institute Reports*: E.A. High Commission, Nairobi.

HALLORAN, PATRICIA O'CONNOR (1955): 'A Bibliography of References to Diseases in Wild Mammals and Birds': *Am. J. vet. Res.*, **16**, 1–465.

LAPIN, B. A. and YAKOVLEVA, L. A. (1963): *Comparative Pathology in Monkeys*: Springfield, Ill., Thomas.

LOMBARD, C. (1962): *Cancérologie comparée*: Paris, Doin.

MEIER, H. (1963): *Epizootiology of Cancer in Animals*: New York, New York Academy of Science.

NEGRONI, P. (1965): *Histoplasmosis*: Springfield, Ill., Thomas.

PICKERING, D. E. (1962) [Ed.]: *Proceedings of a Conference on 'Research with Primates'*: Beaverton, Oregon, Tektronix Foundation.

RUCH, T. C. (1959): *Diseases of Laboratory Primates*: Philadelphia, W. B. Saunders.

STILES, C. W. and HASSELL, A. (1929): 'Key-Catalogue of Parasites Reported for Primates (Monkeys and Lemurs) with their Possible Public Health Importance': Washington, US Treasury Dept., Public Health Service (*U.S. Hyg. Lab. Bull.* (152)), 4.

TASHJIAN, R. J. (1962): *Proceedings: International Symposium on Comparative Medicine*: New York, Eaton Laboratories, Norwich Pharmacol. Company.

VAGTBORG, H. (1964 [Ed.]: *The Baboon—An Annotated Bibliography*: San Antonio, Texas, Southwest Foundation for Research and Education.

VAGTBORG, H. (1965) [Ed.]: *The Baboon in Medical Research*: Austin, Texas, University of Texas.

VANBREUSEGHEM, (1958): *Mycoses of Man and Animals*: London, Pitman.

WHITELOCK, O. v. ST. (1960) [Ed.]: 'Care and Diseases of the Research Monkey': New York, *Ann. N.Y. Acad. Sci.*

ZDRODOVSKII, P. F. and GOLINEVICH, E. H. (1960): *The Rickettsial Diseases*: Pergamon, New York, Oxford, London, Paris.

Author Index

Subject Index